GW00502917

WELCOME!

You hold in your hands the latest and greatest edition of The Official Raspberry Pi Handbook, absolutely crammed with everything we could fit into the pages of this book, covering everything from an incredible ceiling orrery that tracks the motion of the planets to electronic starter kits for Raspberry Pi Pico.

Speaking of Pico, the incredible microcontroller has been upgraded so it now includes a wireless chip. It was great before but now it can natively get data from the internet or over your network for some amazing IoT projects. We've got a guide to what's new, and how to use it for your electronic project ideas.

We've carefully selected projects to show off the broad range of uses Raspberry Pi and Pico can have, whether you're just starting out with a little coding, or looking for your next big project. I believe something in here will truly inspire you make something wonderful.

Happy making, folks.

Rob Zwetsloot

FIND US ONLINE magpi.cc **GET IN TOUCH** magpi@raspberrypi.com

The MagPi

Available on the App Store
GET IT ON Google Play

EDITORIAL
Editor: **Lucy Hattersley**
Features Editor: **Rob Zwetsloot**
Contributors: **David Crookes, PJ Evans, Rosie Hattersley, Phil King, Nicola King, KG Orphanides, Nik Rawlinson, Mark Vanstone**

DISTRIBUTION
Seymour Distribution Ltd
2 East Poultry Ave, London, EC1A 9PT | **+44 (0)207 429 4000**

DESIGN
Critical Media: **criticalmedia.co.uk**
Head of Design: **Lee Allen**
Designers: **Olivia Mitchell, Sam Ribbits**
Illustrator: **Sam Alder**

MAGAZINE SUBSCRIPTIONS
Unit 6, The Enterprise Centre, Kelvin Lane, Manor Royal, Crawley, West Sussex, RH10 9PE | **+44 (0)207 429 4000**
magpi.cc/subscribe
magpi@subscriptionhelpline.co.uk

PUBLISHING
Publishing Director: **Russell Barnes**
russell@raspberrypi.com

Advertising: **Charlotte Milligan**
charlotte.milligan@raspberrypi.com
Tel: +44 (0)7725 368887

Director of Communications: **Liz Upton**
CEO: **Eben Upton**

CC BY NC SA

recycle — When you have finished with this magazine please recycle it

This bookazine is printed on paper sourced from sustainable forests and the printer operates an environmental management system which has been assessed as conforming to ISO 14001.

This official product is published by Raspberry Pi Ltd, Maurice Wilkes Building, Cambridge, CB4 0DS. The publisher, editor and contributors accept no responsibility in respect of any omissions or errors relating to goods, products or services referred to or advertised in the magazine. Except where otherwise noted, content in this magazine is licensed under a Creative Commons Attribution-NonCommercial-ShareAlike 3.0 Unported (CC BY-NC-SA 3.0). ISSN: 978-1-912047-42-0.

Contents

Project Showcases

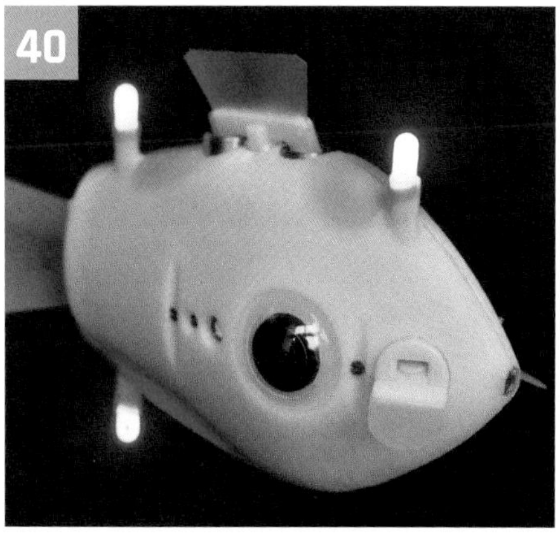

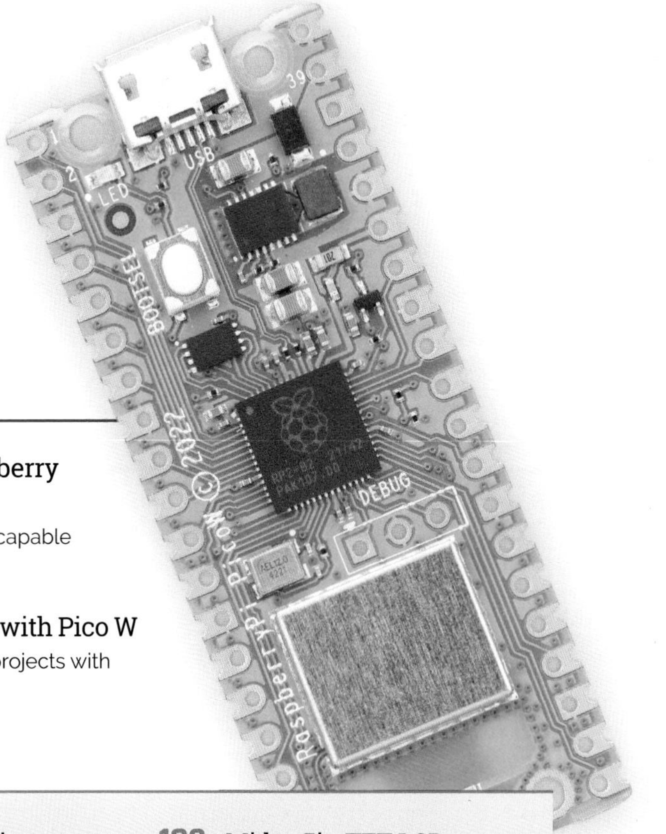

Maker Guide

Raspberry Pi Pico

Reviews

GET STARTED WITH RASPBERRY PI

Set up your Raspberry Pi computer with Raspberry Pi OS, the latest version of the operating system and discover all the new features. By **Phil King**

Whichever model of Raspberry Pi you have, it is part of the most creative computer family on Earth. With a Raspberry Pi, you can hack, make, and build all kinds of different things. It could be a digital camera, a retro games console, or a home media centre. You could even control a sensor on board the International Space Station if you submit an entry for the Astro Pi missions (**astro-pi.org**).

In this guide, we show you how to get started with Raspberry Pi using the new Debian 'Bullseye' edition of Raspberry Pi OS, as well as connecting and controlling some basic electronics.

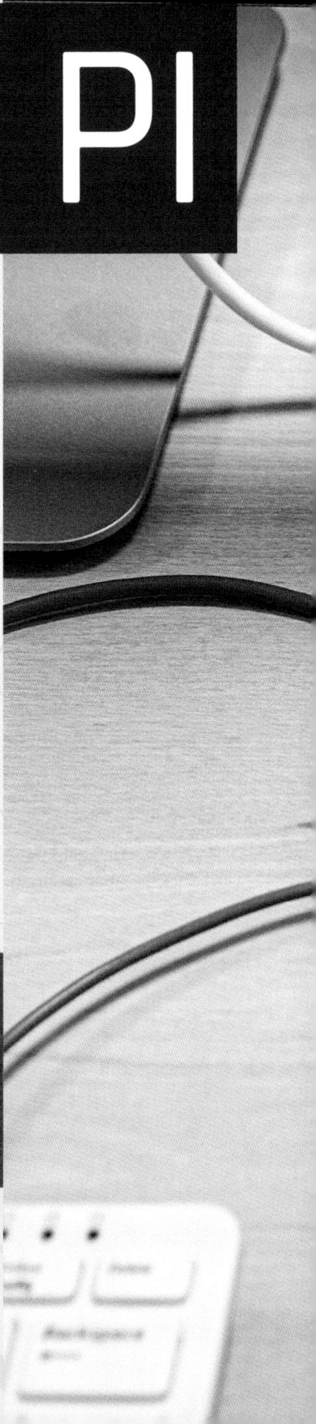

USING RASPBERRY PI OS (BULLSEYE)

Explore the latest version of the default operating system

To make a Raspberry Pi work, you'll need to install an operating system. Unless your Raspberry Pi came with one already preloaded onto a microSD card, you'll have to write the OS to the card.

While other operating systems are available for Raspberry Pi (check out our feature in *The MagPi* #111, **magpi.cc/111**), the official one is Raspberry Pi OS, which has recently been updated to the 'Bullseye' version of Debian Linux.

It's easy to install Raspberry Pi OS onto a microSD card using the Raspberry Pi Imager tool – download it on another computer (Windows, Mac, or Linux) from **magpi.cc/imager**. See the 'Installing Raspberry Pi OS' box for details.

> "You'll need to install an operating system"

Upon inserting the microSD card with the OS on it into your Raspberry Pi's slot and powering up, it will first expand the file system before booting to the Raspberry Pi OS desktop. The Welcome to Raspberry Pi wizard will take you through configuration options including language and time zone, prompt you to change the default password, ask whether all of the taskbar fits on the screen, and prompt you to connect to your wireless router by entering its password.

You'll also be asked if you want to check for and install any software updates, which you can do or skip for now. With setup complete, you'll be prompted to hit Restart to reboot your Raspberry Pi. This time it'll boot straight to the desktop, without the wizard, and be ready to use.

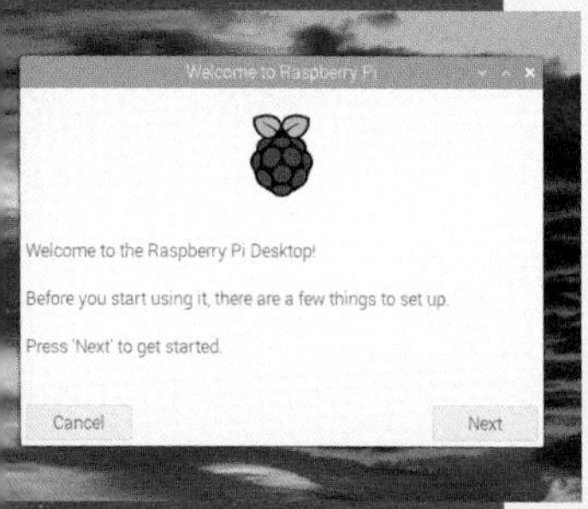

▲ Upon first booting up Raspberry Pi OS, a welcome wizard will guide you through some configuration options

Explore the desktop

Like most operating systems, the standard version of Raspberry Pi OS comes with a desktop interface that you can navigate with a connected mouse.

The default web browser in Raspberry Pi OS is Chromium, although you can install others such as Firefox ESR, Midori, Vivaldi, and Puffin.

Chromium is the basis for Google Chrome, so you may well find its user interface very familiar. There's an Omnibox where you can enter web addresses or search terms. The default search engine is DuckDuckGo, which is focused on preserving the privacy of searchers, but can be changed in the Settings (after clicking on the three dots icon in the top right).

One drawback is that you can't sync your Google account in Chromium to use the same bookmarks and settings that you have in Chrome; Firefox is an alternative browser that enables cross-platform syncing. Other than that, most features are present, including the ability to add extensions from the Chrome Web Store and also to group tabs together. You can also install web apps for some sites such as YouTube, by clicking the option that appears on the right side of the Omnibox.

Install extra software

The standard Raspberry Pi OS only comes with a handful of core applications pre-installed – although there is a 'Full' version of the OS supplied with a lot more software (find it in Raspberry Pi Imager, under 'Raspberry Pi OS (other)').

It's simple to install any extra software you want, however. By far the easiest way is to use the Recommended Software tool (Menu > Preferences > Recommended Software). You can then browse a range of applications; to install one, simply tick its box and click Apply.

Applications in Recommended Software include the Claws email client and the LibreOffice productivity suite. The latter features six applications: the Writer word processor, Calc spreadsheet, Impress presentation, Draw diagrams, Base database, and Math formula editor. It can load/save Microsoft Office documents too.

If you can't find what you need in Recommended Software, you will be able to install additional software packages using the Add/Remove Software tool, or by entering commands (such as `sudo apt install` and `sudo pip install`) in a Terminal window.

GET TO KNOW RASPBERRY PI OS' INTERFACE

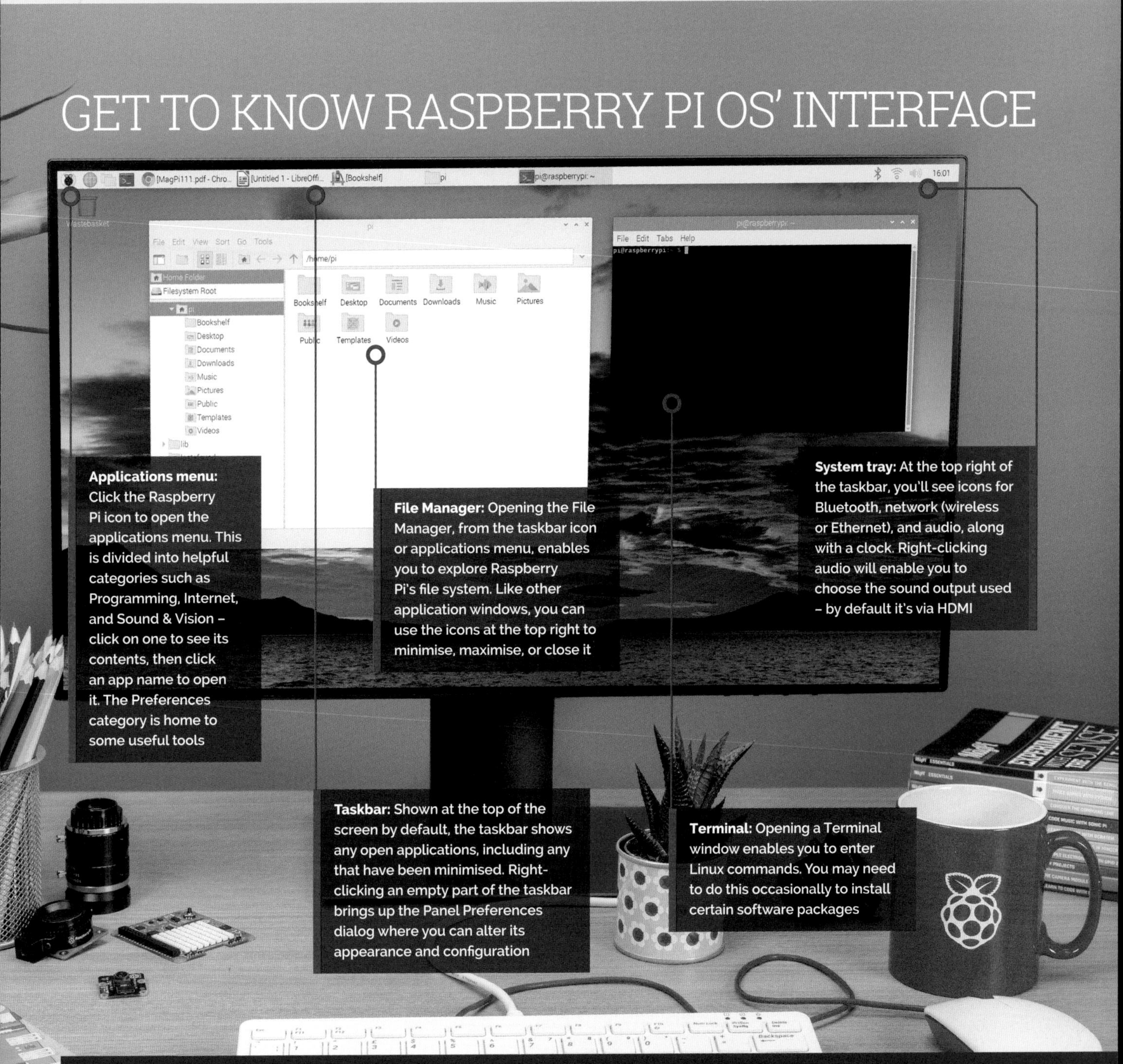

Applications menu: Click the Raspberry Pi icon to open the applications menu. This is divided into helpful categories such as Programming, Internet, and Sound & Vision – click on one to see its contents, then click an app name to open it. The Preferences category is home to some useful tools

File Manager: Opening the File Manager, from the taskbar icon or applications menu, enables you to explore Raspberry Pi's file system. Like other application windows, you can use the icons at the top right to minimise, maximise, or close it

System tray: At the top right of the taskbar, you'll see icons for Bluetooth, network (wireless or Ethernet), and audio, along with a clock. Right-clicking audio will enable you to choose the sound output used – by default it's via HDMI

Taskbar: Shown at the top of the screen by default, the taskbar shows any open applications, including any that have been minimised. Right-clicking an empty part of the taskbar brings up the Panel Preferences dialog where you can alter its appearance and configuration

Terminal: Opening a Terminal window enables you to enter Linux commands. You may need to do this occasionally to install certain software packages

Installing Raspberry Pi OS

To install Raspberry Pi OS (or upgrade from an earlier version of Raspberry Pi OS to the latest 'Bullseye' edition), you will need to install a fresh version of the OS to your microSD card. Typically you'll do this using Raspberry Pi Imager (**magpi.cc/imager**) on a Windows, Apple Mac, or another Linux computer (including another Raspberry Pi). Open Raspberry Pi Imager on your other computer and insert your microSD card (using a USB adapter if needed). Click 'Choose OS' and select 'Raspberry Pi OS (32-bit)'.

Now click 'Choose Storage' and select your inserted microSD card (which may well be labelled as 'Generic STORAGE DEVICE Media' with its storage capacity). Click Write to download Raspberry Pi OS and copy the software to the microSD card.

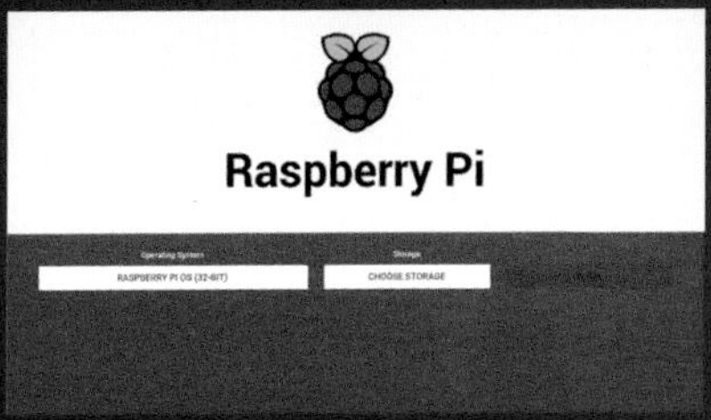

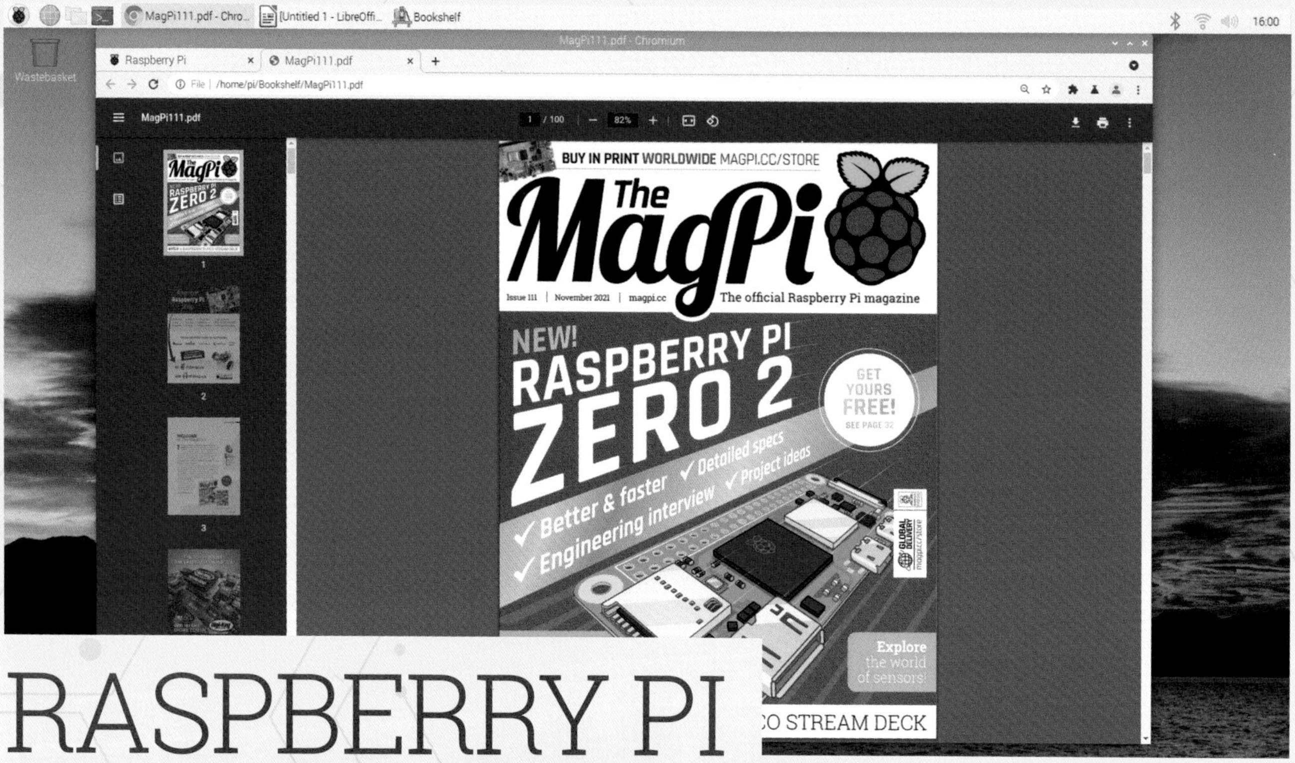

▲ Read digital PDFs of your favourite magazines by downloading issues in the Bookshelf app

RASPBERRY PI OS BULLSEYE FAQ

Find out what's changed in the new version of Raspberry Pi OS

With the basics of the desktop GUI and core applications covered, let's take a look at some of the new features in Raspberry Pi OS Bullseye and the differences from previous versions.

How do I get Bullseye?

Get Raspberry Pi OS Bullseye by installing a fresh installation of the latest version of Raspberry Pi OS from Raspberry Pi Imager (**magpi.cc/imager**).

Can I upgrade from Buster to Bullseye?

You can't upgrade from Debian Buster to Debian Bullseye using `sudo apt full-upgrade` – this only takes you to the latest version of the current (i.e. 'Buster') operating system. You have to install Bullseye onto a fresh microSD card. If you have data on your Buster installation you want to keep, we recommend copying it to a separate drive and then back to the fresh Bullseye installation.

What if I want to get Buster?

Buster is now known as Raspberry Pi OS (Legacy) and is available from Raspberry Pi's Software page (**magpi.cc/buster**). It can also be found in Raspberry Pi Imager. Read more about Raspberry Pi OS (Legacy) in Gordon Hollingworth's blog post (**magpi.cc/legacy**).

How different does Bullseye look?

If you've used Raspberry Pi OS before, you'll immediately notice that the default wallpaper is different in Bullseye. If you like, you can change it in Preferences > Appearance Settings.

Another change you may notice is that desktop windows have a shadow effect on their borders, and animate as they open and close. Widgets and their tabs and buttons also look a little different.

In File Manager, the view modes have been simplified, with icon options to switch between icons or list mode. For more advanced options, you can use the View menu to zoom the icon size in/out and select thumbnail icon mode.

What's new in Bullseye?

One of the numerous under-the-hood changes in Raspberry Pi OS Bullseye is that the KMS (kernel mode setting) video driver is now used by default, whereas previously it was an experimental option.

While you may not notice much of a change in how the video display works or performs in general, one major advantage of using KMS is that it's the standard open-source video driver used in Linux. This means that any application written using the standard Linux display APIs should run on Raspberry Pi without the need for modification.

In addition, now that all the display drivers for Raspberry Pi OS are part of the Linux kernel, this should make it easier for manufacturers of custom displays to add support for Raspberry Pi.

The upgrade to Bullseye brings with it an update of Chromium to version 92, which has also been optimised to use Raspberry Pi's hardware to accelerate video playback.

> Bullseye takes its name from a character in Disney's Toy Story movie franchise

Anything else new?

The Bookshelf application now includes free downloadable PDFs of our sister publication Custom PC, the UK's best-selling magazine for PC hardware, overclocking, gaming, and modding. Bookshelf also contains issues of *The MagPi*, HackSpace, and Wireframe magazines, along with our official Raspberry Pi books.

Will my Raspberry Pi run faster?

There's a speed boost bonus for owners of recent Raspberry Pi 4 and Raspberry Pi 400 models: the default turbo-mode clock has increased from 1.5GHz to 1.8GHz. For more details, see Eben Upton's blog post: **magpi.cc/bullseyeboost**. You can manually overclock many models of Raspberry Pi (**magpi.cc/overclock**).

Why doesn't my Raspberry Pi have the new interface?

The new look is down to the use of a new window manager called Mutter (**magpi.cc/mutter**). The switch to Mutter is due to the upgrade from GTK+2 to GTK+3 for user interface components. Mutter replaces the Openbox window manager used in previous versions of the OS. Openbox is still employed if your Raspberry Pi model has less than 2GB of RAM, because Mutter is quite demanding in terms of memory usage.

▲ Playing a video in the VLC application; it runs very smoothly in Bullseye

How do I manage notifications?

Raspberry Pi OS Bullseye now has a common notification manager which tells you about things like USB devices being inserted/removed and low power supply voltage, and memory allocation.

Notifications appear in grey windows under the right end of the taskbar, with the most recent messages shown at the top. Each notification will time out and automatically disappear after 15 seconds, or when you click on it.

The notification timeout setting can be altered in the Appearance tab of the Panel Preferences dialog, accessed by right-clicking the taskbar. If you set the timeout to zero, notifications will stay on screen until clicked. You can also turn notifications off completely (not advisable).

How should I keep Bullseye up-to-date?

Another major user-friendliness improvement in Bullseye is the introduction of an updater plug-in for the taskbar. Whenever you reboot Raspberry Pi, or every 24 hours if it's left powered up, it will check for any software updates and show an update icon in the taskbar with a notification telling you that updates are available.

Click the update notification to open a window showing you the packages that can be updated. Just click the Install button to start installing them; you can then carry on with whatever you were doing while they install in the background.

Why is it 'Bullseye'?

Every major version of the Debian Linux distro, upon which Raspberry Pi OS is based, takes its name from a character in Disney's *Toy Story* movie franchise. Previous versions include Wheezy, Jessie, Stretch, and Buster. If you've not seen the movies, Bullseye is Woody and Jessie's horse, one of the new characters introduced in *Toy Story* 2.

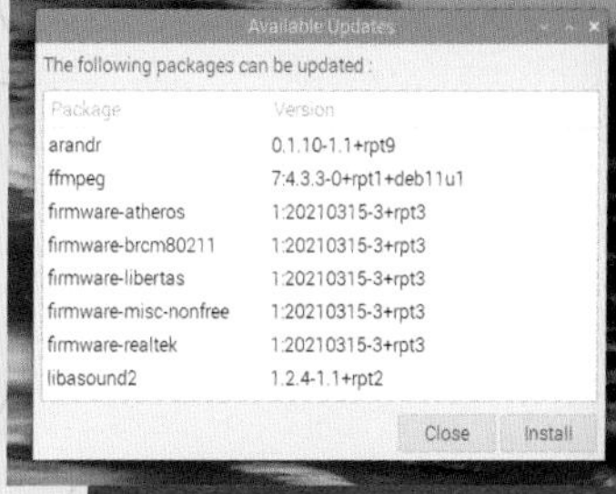

▲ Clicking the updater icon in the taskbar brings up a list of available software updates

BULLSEYE & RASPBERRY PI **CAMERA SYSTEM**

Get to grips with using Raspberry Pi cameras with Bullseye! By **David Plowman**

When we released our first Raspberry Pi OS image based on Debian Bullseye, we pointed to a change that is hugely important to people who have written code to use cameras with Raspberry Pi: the driver that Raspberry Pi uses to access Camera Modules has been replaced with libcamera.

These very significant changes mean less closed-source code, and they make it easier for people outside of Raspberry Pi to develop new camera hardware and software; but they also mean that new Raspberry Pi OS releases will no longer support the familiar raspicam apps and picamera Python library.

In the place of this older camera system is the new and almost entirely open-source camera stack based on standard Linux frameworks such as V4L2 (Video for Linux) and libcamera. Our kernel drivers have been moving in this direction for some time too, and have just recently taken further large strides towards the preferred new Media Controller architecture.

But the principal difference that users will notice is that OS releases from Bullseye onwards will no longer support the older camera system and applications, and Raspberry Pi's libcamera-apps will be built and pre-installed instead. Before we go further, note that Raspberry Pi OS Buster is still available to download if you're not ready to use Bullseye. If you are using camera applications with your Raspberry Pi, we recommend you take some time to weigh up whether to move to Bullseye at this point. A blog post on the Raspberry Pi website (**magpi.cc/bullseyecamera**) considers why, and why not, you might want to do so.

Raspberry Pi OS (Legacy)

A version of Raspberry Pi OS based on the older 'Buster' system is available. Read about it on this Raspberry Pi post. **magpi.cc/legacy**

David Plowman is an Engineer at Raspberry Pi

What are libcamera-apps?

Libcamera-apps are designed to copy most of the functionality that users will know from raspistill, raspivid, and raspiyuv. There are some unavoidable differences, which are examined in greater detail on Raspberry Pi's documentation pages (**magpi.cc/libcamdiff**). The new applications include:

libcamera-hello – a simple 'hello world' application that starts a camera preview stream and displays it on the screen.

libcamera-jpeg – a simple application to run a preview window and then capture high-resolution still images.

libcamera-still – a more complex still image capture application that emulates more of the features of raspistill.

libcamera-vid – a video capture application.

libcamera-raw – a basic application for capturing raw (unprocessed Bayer) frames directly from the sensor.

libcamera-detect – this application is not built by default, but users can build it if they have TensorFlow Lite installed on their Raspberry Pi. It captures JPEG images when certain objects are detected.

Bullseye blog

For detailed information on Bullseye, straight from the horse's mouth. Read Simon Long's Bullseye blog post. **magpi.cc/bullseye**

" A new camera stack based on standard Linux frameworks "

Legacy Camera Support

Legacy Camera Support is a new option in Raspberry Pi OS (Bullseye). It replaces the "Enable/disable connection to the Raspberry Pi Camera" found in Raspberry Pi OS (Buster) and previous Raspberry Pi operating systems. (You no longer need to to enable camera support in Rasperry Pi OS (Bullseye) as the libcamera API is enabled by default).

Legacy Camera Support is included in the latest update to Raspberry Pi OS. Open a terminal window and enter:

```
sudo apt-update
sudo apt full-upgrade
```

You will need to reboot Raspberry Pi OS for the changes to take effect (Menu > Shutdown and click Reboot).

Use raspi-config to enable Legacy Camera Mode. In a terminal window enter:

```
sudo raspi-config
```

Chose '3 Interface Options' and 'I1 Legacy Camera' and select 'Yes' in the 'Would you like to enable legacy camera support?' window. You will see a message saying 'Legacy camera support is enabled' along with 'Please note that this functionality is deprecated and will not be supported for future development'. Legacy Camera Mode will enable you to use Picamera-based projects with Raspberry Pi OS (Bullseye), but you should move your project to libcamera for future compatibility.

See the Raspberry Pi Documentation for more information on using the raspi-config tool (**magpi.cc/raspiconfig**).

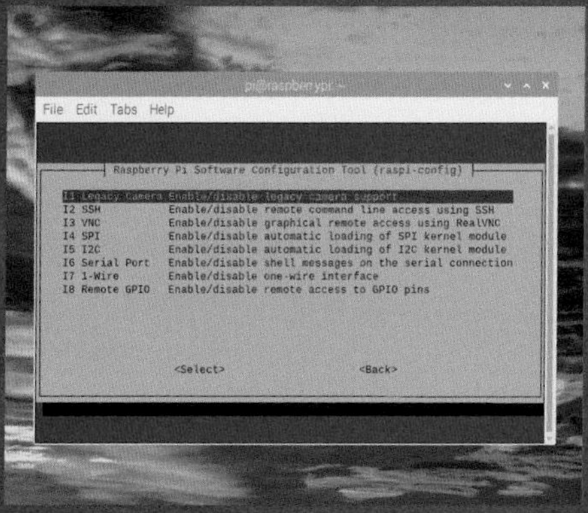

▲ Enable Legacy Camera Mode in raspi-config

CONNECTING **ELECTRONICS**

Make use of Raspberry Pi's GPIO header to connect and control circuits. By **Phil King**

While Raspberry Pi can be used as an effective desktop PC, a major bonus is that it can easily be connected to electronic circuits and add-on boards (often called HATs) via its 40-pin GPIO header. This is found on every Raspberry Pi model, including the tiny Zero 2 W and also the Pi 400. See the GPIO diagram below for details of pin functions.

There's a wide variety of HATs (Hardware Attached on Top) and other expansion boards available for Raspberry Pi. On most Raspberry Pi models, the HAT just fits on top of the GPIO pins and is positioned over the computer's PCB. Since Raspberry Pi 400's GPIO header is located at the rear of the keyboard unit and is slightly inset, however, you may well need a ribbon cable or breakout adapter to connect a HAT.

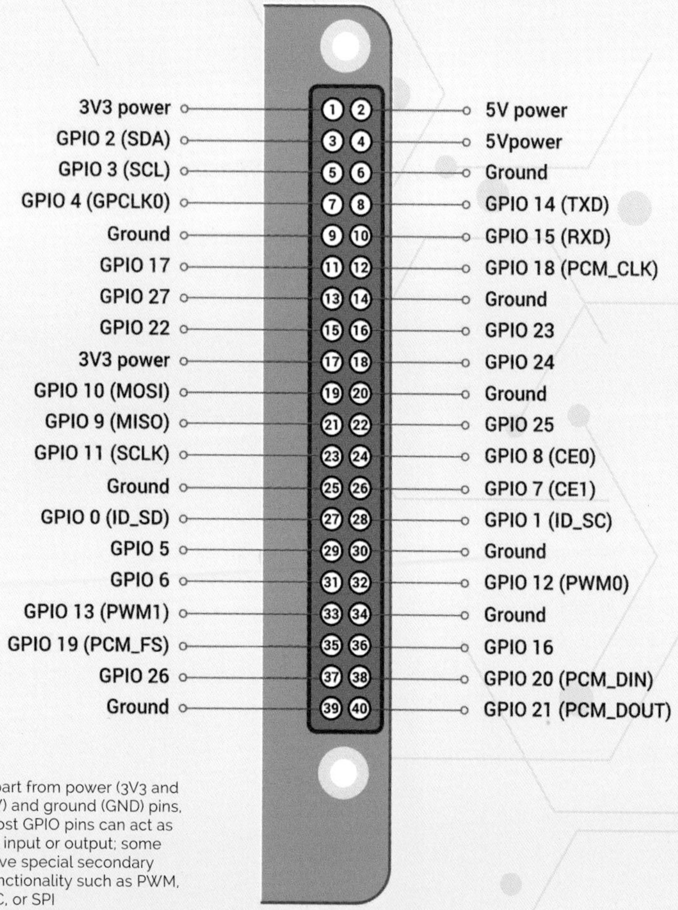
3V3 power
GPIO 2 (SDA)
GPIO 3 (SCL)
GPIO 4 (GPCLK0)
Ground
GPIO 17
GPIO 27
GPIO 22
3V3 power
GPIO 10 (MOSI)
GPIO 9 (MISO)
GPIO 11 (SCLK)
Ground
GPIO 0 (ID_SD)
GPIO 5
GPIO 6
GPIO 13 (PWM1)
GPIO 19 (PCM_FS)
GPIO 26
Ground
5V power
5Vpower
Ground
GPIO 14 (TXD)
GPIO 15 (RXD)
GPIO 18 (PCM_CLK)
Ground
GPIO 23
GPIO 24
Ground
GPIO 25
GPIO 8 (CE0)
GPIO 7 (CE1)
GPIO 1 (ID_SC)
Ground
GPIO 12 (PWM0)
Ground
GPIO 16
GPIO 20 (PCM_DIN)
GPIO 21 (PCM_DOUT)

▶ Apart from power (3V3 and 5V) and ground (GND) pins, most GPIO pins can act as an input or output; some have special secondary functionality such as PWM, I2C, or SPI

Alternatively, connect and control your own electronic circuits to the GPIO pins using jumper wires. Standard electronic components such as LEDs, push-buttons, and sensors can be used, often via a breadboard (**magpi.cc/breadboard**).

JAM HAT

For this introductory electronics example we'll make things simpler by using a JAM HAT. Available from The Pi Hut (**magpi.cc/jamhat**), this add-on board features built-in LEDs, push-buttons, and a piezo buzzer. Just slot it onto your Raspberry Pi's GPIO header (or connect to Pi 400 via a ribbon cable or breakout adapter).

> " Set up the LEDs as traffic lights and use a push-button for a pedestrian crossing "

You are now ready to start programming your first electronics project in the popular Python language. In Raspberry Pi OS, open the Thonny IDE (Menu > Programming > Thonny IDE). The easiest way to read and control the GPIO pins in Python is to use the GPIO Zero library. This even has a special class for the JAM HAT – at the start of our code, we import this with: `from gpiozero import JamHat`.

So that we can incorporate a delay in our program, we also import the `sleep` method from the time library: `from time import sleep`.

We then assign a variable, `jamhat`, to the `JamHat` class, with: `jamhat = JamHat()`.

Now, to blink the two rows (`lights_1` and `lights_2`) of LEDs in turn, we can use GPIO Zero's `blink` method:

```
jamhat.lights_1.blink()
sleep(0.5)
jamhat.lights_2.blink()
```

See the **jam-LEDs.py** listing for the full code. Run it in Thonny to see the effect. We can specify which colour LED in each row to blink. To turn on the red LED in row 1 for one second enter:

WIZnet

"

How to Add Ethernet for Raspberry Pi Pico or RP2040

at maker.wiznet.io/contest

"

WIZnet Ethernet Hat Contest Winners

AI-driven IoT 3D Printer Motion & Status Tracker w/ Telegram

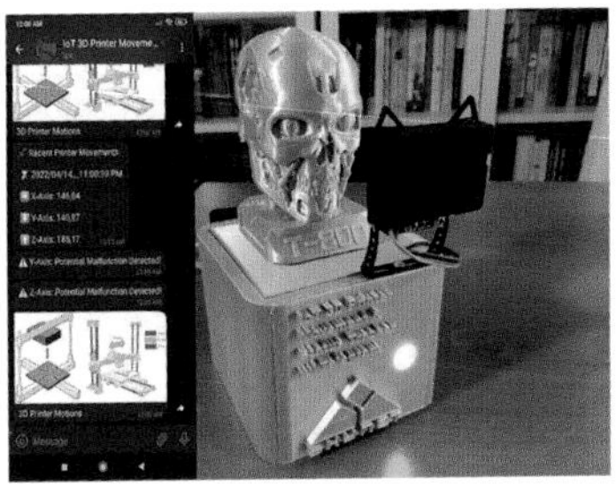

This project is one of WIZnet's Etherernet HAT Contest 2022 award-winning projects that tracks vertical and lateral moverments of a 3D printer via AirTags.

It allows users to get informed of malfunctions related to motion via Telegram. It's a great piece of IoT 3D Printer Motion & Status built from an AI Camera, Raspberry-Pi-Pico ,and WIZnet's Ethernet HAT. Its detailed step by step information is easy to follow and available.

Ethernet CAN Gateway

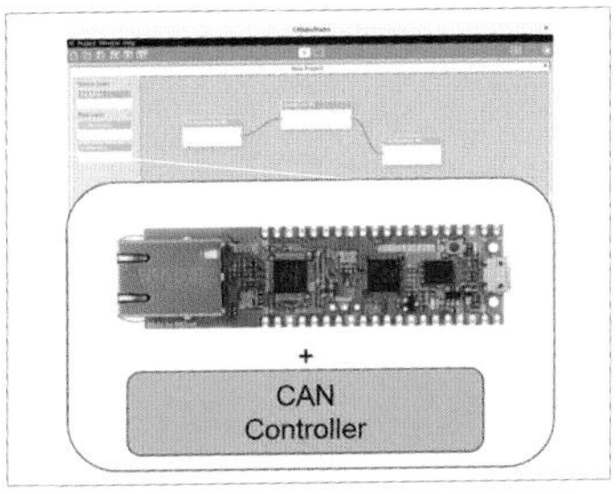

If you are interested in communication used in cars, you should consider the CAN communication device. This project is an Ethernet to CAN Gateway project, utilizing the W5100S-EVB-Pico and the CAN module MCP2515. This project is also detailed in a Step-by-Step process, from PCB to Firmware Source/Case, making it easy to follow for those who make Ethernet to CAN Gateway.

Rotary Dial Remote Control for Home Automation

Don't you remember the rotary dial classic phones?

This project converts the classic phone to a Remote Control and Alarm for Home Automation. If you are looking for a classic feel for a Home Automation System, follow the link below.

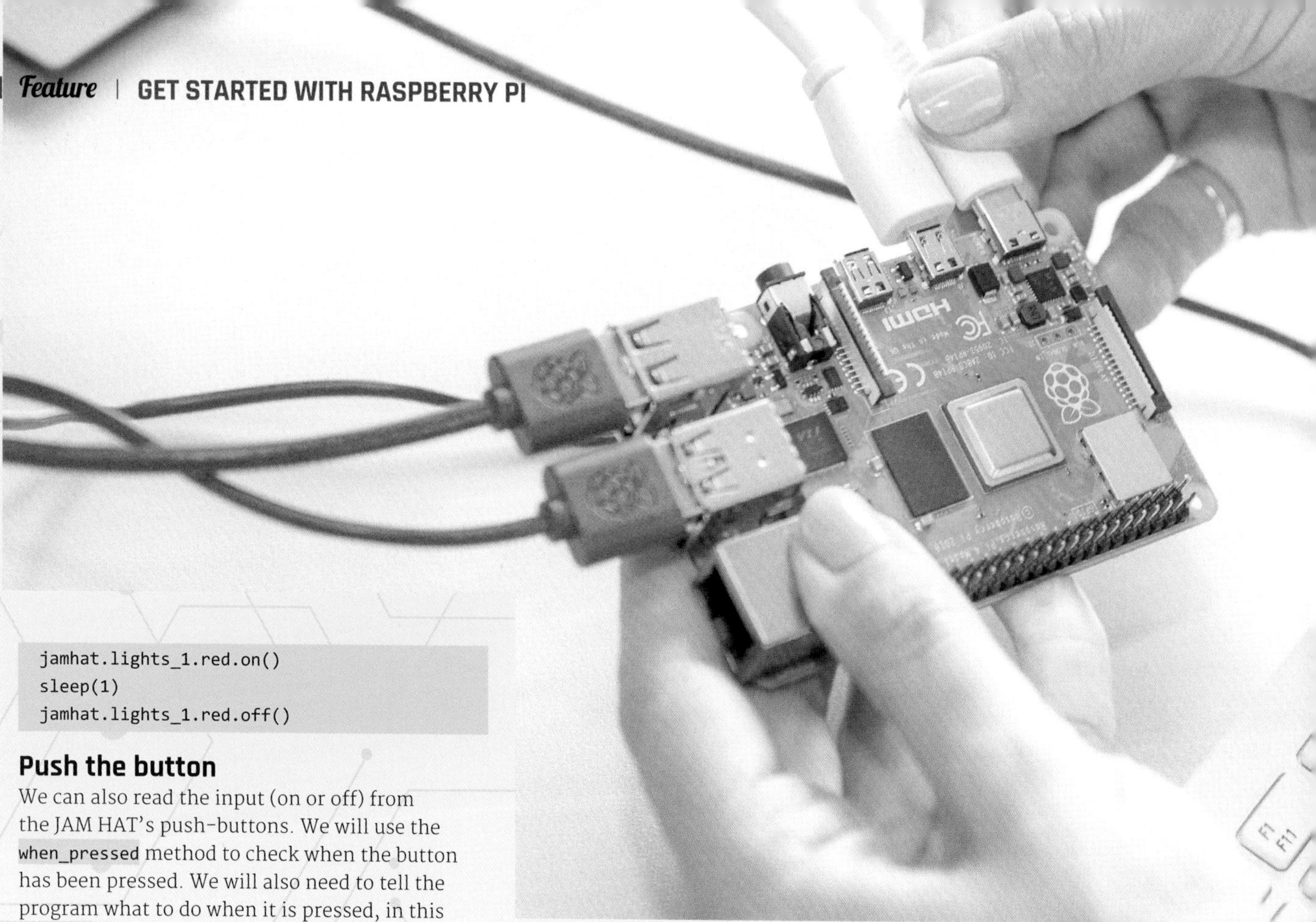

```
jamhat.lights_1.red.on()
sleep(1)
jamhat.lights_1.red.off()
```

Push the button

We can also read the input (on or off) from the JAM HAT's push-buttons. We will use the `when_pressed` method to check when the button has been pressed. We will also need to tell the program what to do when it is pressed, in this case turning on all the JAM HAT's LEDs and its buzzer using the jamhat.on command:

```
jamhat.button_1.when_pressed() = jamhat.on()
```

See the **jam-button.py** listing for the full code. Run it and every time you press button 1, the LEDs will light up and the buzzer will make a noise. Press button 2 to turn them off.

There's a lot you can do with the JAM HAT, including setting up the LEDs as traffic lights and using a push-button for a pedestrian crossing. For more details, see the JAM HAT getting-started guide (**magpi.cc/jamhatguide**) and further code on GitHub (**magpi.cc/jamhatgit**).

▼ Making electronics easier, the JAM HAT features six coloured LEDs, two tactile push-buttons, and a buzzer

jam-LEDs.py

> Language: **Python 3**

```
from gpiozero import JamHat
from time import sleep

jamhat = JamHat()

jamhat.lights_1.blink()
sleep(0.5)
jamhat.lights_2.blink()
```

jam-buttons.py

> Language: **Python 3**

```
from gpiozero import JamHat

jamhat = JamHat()

jamhat.button_1.when_pressed =
jamhat.on
jamhat.button_2.when_pressed =
jamhat.off
```

Eval Boards based on Raspberry Pi RP2040

WIZnet

W5100S-EVB-Pico

$9.95

aws Qualified Device AWS IoT Core

Microsoft Azure Certified Device

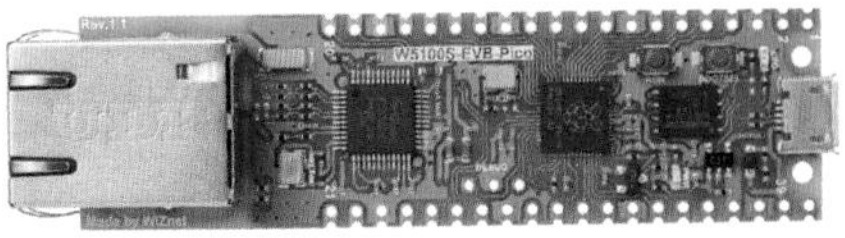

W5100S

RP2040

W5100S integrates the hardwired TCP/IP stack with 10/100 Ethernet MAC and PHY - truly a one-chip solution for stable internet connectivity.

W5500-EVB-Pico

$9.95

aws Qualified Device AWS IoT Core

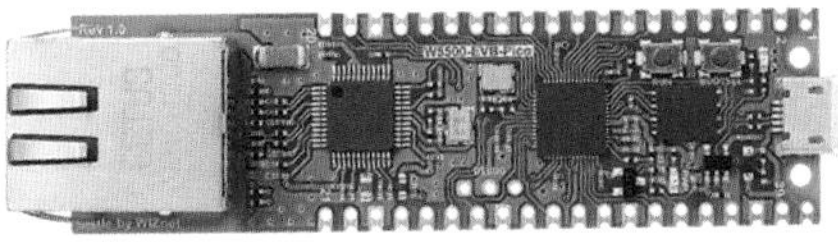

W5500

RP2040

W5500 provides 8 independent SOCKETs to be used simultaneously and 32KB internal memory for data communication.

WIZnet Ethernet HAT

$4.95

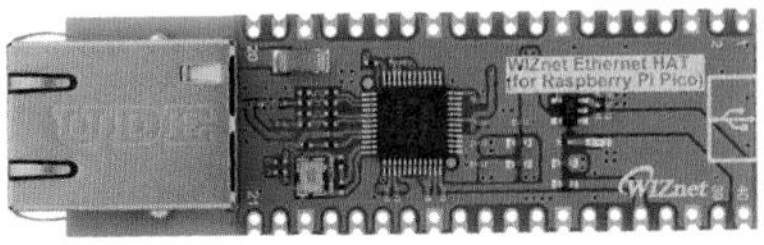

W5100S

WIZnet Ethernet HAT (Hardware Attached on Top) is a Raspberry Pi Pico pin-compatible board that utilizes W5100S and supports both 3.3V & 5V.

WizFi360-EVB-Pico

$6.95

aws Qualified Device AWS IoT Core

WizFi360

RP2040

WizFi360 is a low cost and low-power consumption industrial-grade WiFi module, compatible with IEEE802.11 b/g/n standard and supports SoftAP, Station and SoftAP+Station modes.

WIZnet Online shop eshop.wiznet.io

WIZnet Docs docs.wiznet.io

Github github.com/wiznet

Introducing Raspberry Pi Zero 2 W

The smallest Raspberry Pi now packs a quad-core processor and runs over five times faster. Raspberry Pi Zero is back in a big way.

By **Lucy Hattersley**

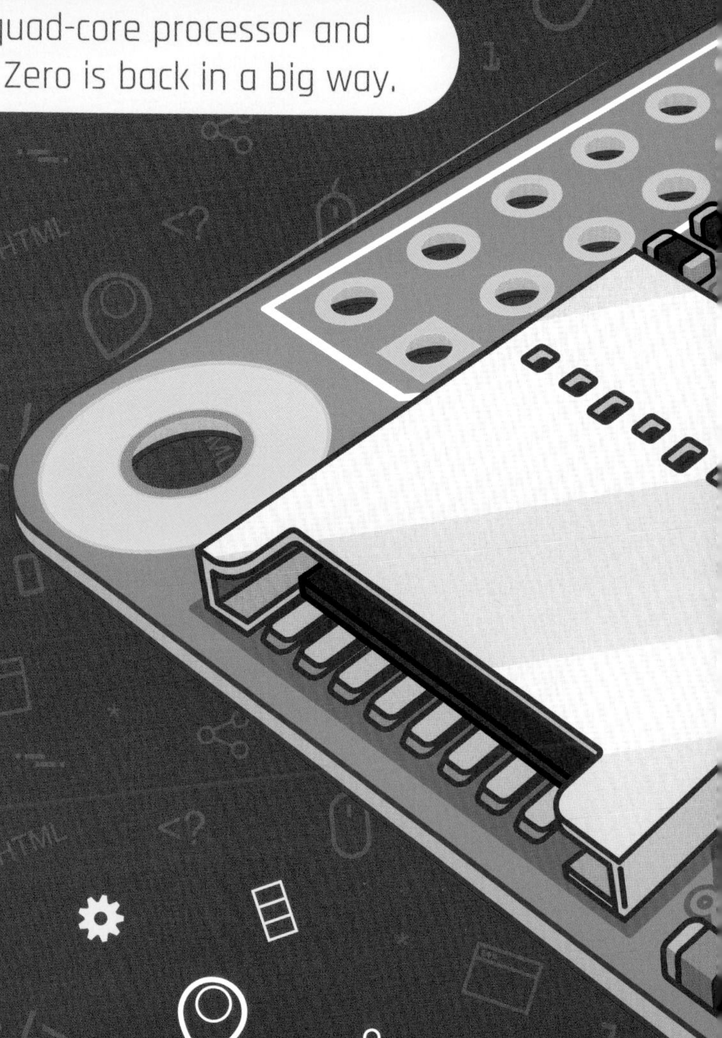

Raspberry Pi Zero is one of the greatest ever Raspberry Pi computers. This tiny powerhouse has been close to the heart of all *The MagPi* magazine readers ever since its introduction in 2015.

It may be diminutive in size, but Raspberry Pi Zero's reputation is enormous. In this book we're delighted to introduce you all to Raspberry Pi Zero 2 W: it has a lot to live up to.

It comes in the exact same package you know and love. Only now, Zero 2 packs a more powerful Arm Cortex-A53 quad-core processor running at 1GHz.

Zero 2 sticks with the same form factor, so you can take a Raspberry Pi Zero out of your current project, and drop Raspberry Pi Zero 2 W straight in its place and immediately benefit from the improved speed. It's also compatible with the vast array of kits and innovative projects designed around the small board.

At the heart of Raspberry Pi Zero 2 W sits an all-new SiP (System-in-Package). An exciting new approach by Raspberry Pi, that combines the usual System-on-Chip with DRAM and enables a faster CPU to sit in the same form factor.

It's a new, and faster, day for Raspberry Pi Zero, yet one that keeps the important heritage intact. We can't wait to see what you make with it.

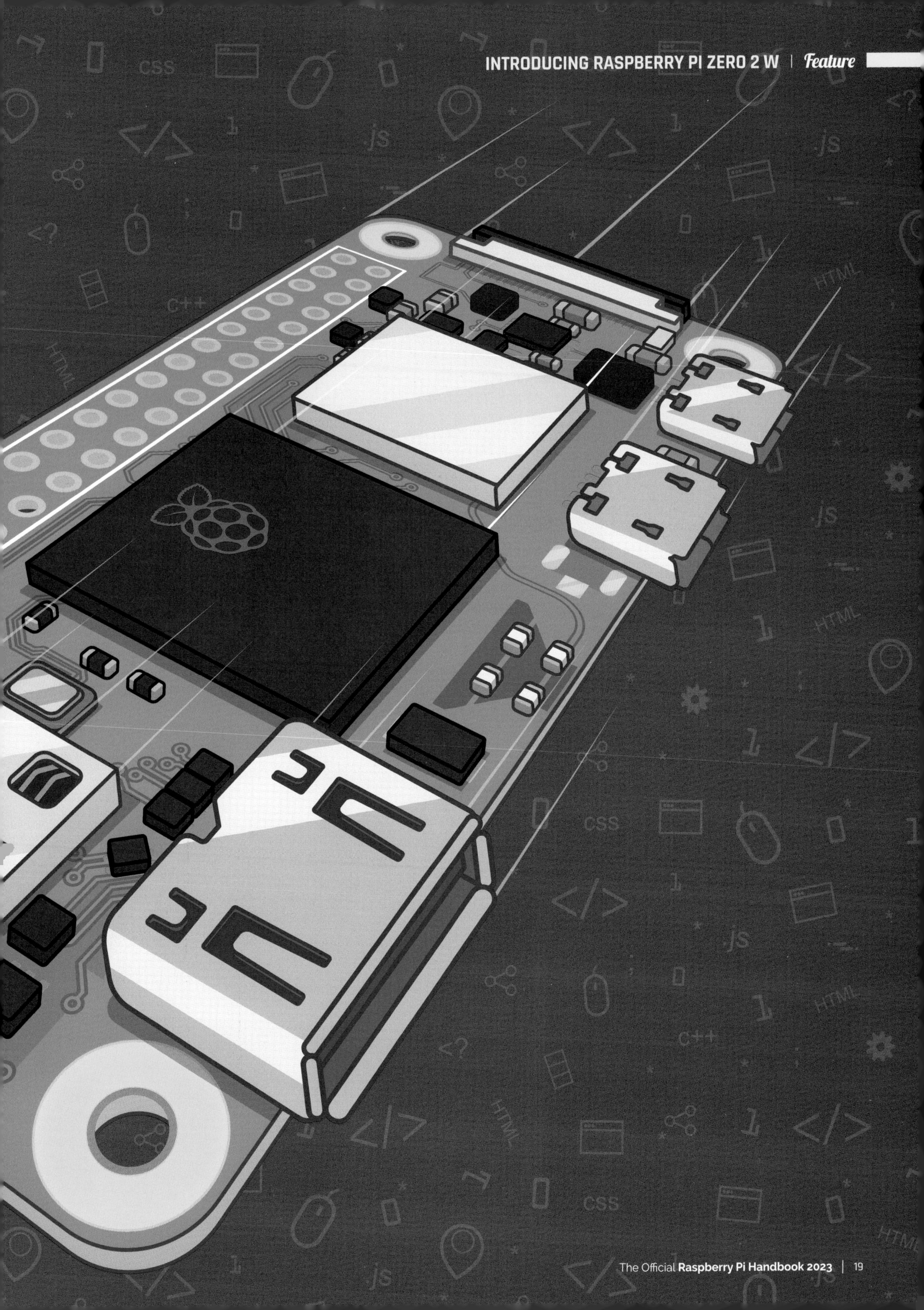

Meet Raspberry Pi Zero 2 W

Raspberry Pi Zero 2 W is designed to be instantly swappable with Raspberry Pi Zero in a current project, with an identical form factor and most components in the same place. Dig a little deeper, and there are a host of tweaks and improvements across the board.

1 MINI-HDMI
The Mini-HDMI port now has winged edging that enables easier insertion of the HDMI cable.

2 SYSTEM-IN-PACKAGE
At the heart of Raspberry Pi Zero 2 sits a System-in-Package (SiP) containing a Broadcom BCM2710A1 quad-core Arm Cortex-A53 System-on-Chip (SoC) with VideoCore IV and 512MB DRAM.

6 ANALOGUE VIDEO AND RESET
Raspberry Pi Zero 2 W supports analogue video and reset connections. The pins are moved to the rear of the board, alongside other test pins. In the bottom-right sits TV, while in the top-left, a Run pin can be used to reset and restart Zero 2.

Specifications

SOC:
Broadcom BCM2710A1 quad-core Arm Cortex-A53 (ARMv8-A) 64-bit @ 1GHz

GPU:
Broadcom VideoCore IV

RAM:
512MB DRAM

NETWORKING:
802.11 b/g/n wireless LAN, Bluetooth 4.2 (Bluetooth Low Energy, BLE)

GPIO:
HAT-compatible 40-pin GPIO header, unpopulated

STORAGE:
microSD

PORTS:
Micro SDCard slot, Mini HDMI, USB On-The-Go, micro USB power; composite video and reset pins (via solder test points), CSI camera connector

POWER CONSUMPTION:
1 W (idle in Raspberry Pi OS GUI); Max power, 3.8 W or 0.76 A (32-bit mode); 5.5 W or 1.1 A (64-bit) mode

PRICE:
$15

DIMENSIONS:
66 × 30.5 × 5 mm form-factor (compatible with Raspberry Pi Zero)

3 NEW POWER CIRCUIT
The enhanced power circuitry from Raspberry Pi 3 has been shrunk down and fitted onto Raspberry Pi Zero 2 W's board.

4 WIRELESS LAN
The wireless LAN and Bluetooth is now housed inside a module enclosure. This can help with compliance, enabling Raspberry Pi Zero 2 W to have a smoother entry to highly regulated industrial environments.

3
4
PWR IN

USB

5 CAPACITORS INSIDE CPU
There are now 18 capacitors inside the SiP and fewer 0201 capacitors on the board. When the CPU demands a surge in power, the capacitors inside can ensure that power is there to meet demand.

Meet the System in Package

One interesting new aspect of Raspberry Pi Zero 2 W is the System-in-Package approach to the chipset design.

A Broadcom BC2710A1 SoC is placed on the substrate and 512MB of DRAM is placed above, with an interposer placed between the two (to act as a 'spacer'). This enables both chips to connect neatly to the substrate using gold wire. The whole thing is placed inside a mould. This is the black chip you see on the Raspberry Pi board.

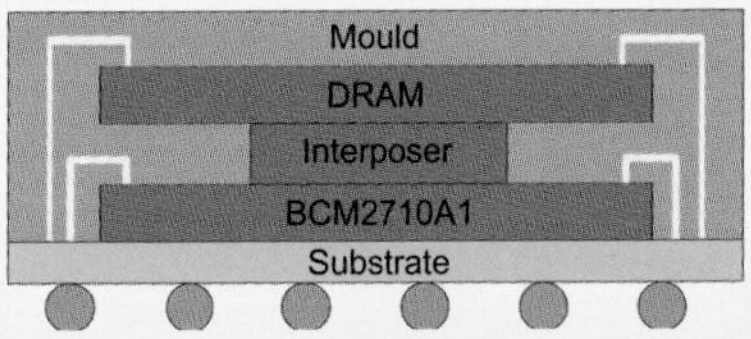
Mould
DRAM
Interposer
BCM2710A1
Substrate

X-ray images

The picture is a view of a 3 mm × 2 mm section of the System-in-Package. The black circles are solder balls. The thick grey lines are tracks on the substrate. The thin black lines are gold wires, and you can see them bonding to the silicon (which is transparent to the X-ray machine). Here, we can see the 0201 (50 um × 25 um in size) capacitors inside the package.

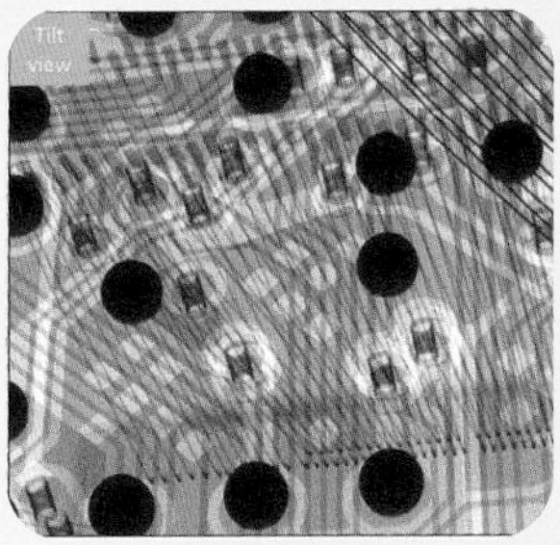

Here is a top-down X-ray of the System-in-Package that reveals an Easter egg! Spot the Raspberry Pi logo shape in the ball map. This is possible because, while the connections on the outside are signals, the ones in the middle are ground/power and can be arranged into any pattern.

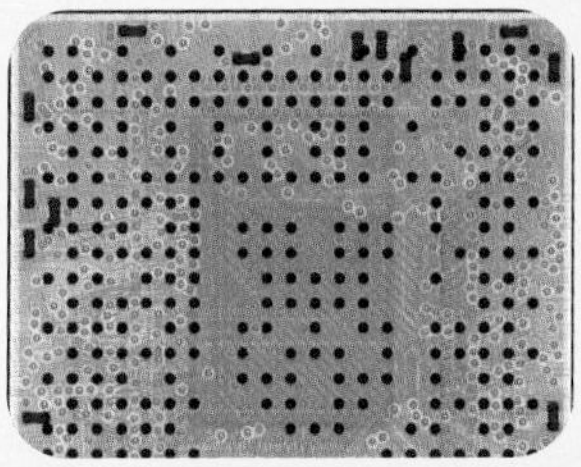

Eben Upton and Simon Martin on Raspberry Pi Zero 2 W

Raspberry Pi founder and CEO Eben Upton, and Principal Hardware Engineer **Simon Martin**, walk us through the design of the new Zero 2

We caught up with Simon Martin, Principal Hardware Engineer at Raspberry Pi and Eben Upton, founder and CEO of Raspberry Pi.

Simon: "Raspberry Pi Zero 2 W is all about how much power you can pack into such a tiny space. It's about just how much can we get from such a small form factor.

"Keeping that same form factor is important. An original Raspberry Pi Zero can be removed from a project and Raspberry Pi Zero 2 W can be plugged into its place. Any application that was there beforehand will get a boost from the processor performance.

"Instead of using the single-core processor that's in the Raspberry Pi Zero, we've got a quad-core Cortex-A53 processor, which is similar silicon to the original Raspberry Pi 3 when that was announced."

Eben: "There's this whole question of 'can we do it again?' You can use Moore's Law in two ways: you can take an amount of money and fill [Raspberry Pi] with more and more computing power, or you can take the current amount of computing power and deliver that at a lower and lower cost. And that's what we did with Raspberry Pi 1 – we took a ten-year-old PC's processing power and delivered it at around $30. And then we launched Raspberry Pi 2 in 2015 and, lo and behold, we've done the thing that everyone else does: we've picked a price point and filled it out with six times as much computing power using Moore's Law.

"Raspberry Pi Zero is really the result of going 'Aha! We should do the same thing as we did before.' We should take a Raspberry Pi 1's worth of computing power and we should use Moore's Law to squeeze the price down.

"Every part on that board pays for itself. It's a single-sided board, and it only has reflowable components; it doesn't have any through-hole components. So it can all be manufactured simply and robotically.

"So, coming up to the sixth anniversary, and how to use everything we've learned from those five years, there's this whole question of 'can we do it again?' And we're delivering most of Raspberry Pi's 3 power for not much more than half the price of a Raspberry Pi 3.

"The reality is that earlier this year people found out that Raspberry Pi knows how to make their own silicon [see, Raspberry Pi RP2040 in Raspberry Pi Pico, **magpi.cc/102** – Ed]. Raspberry Pi Zero 2 W is not our own silicon, but we package silicon in unusual ways

Inside the package

Simon: "It's also known as a 'stacked package'. It's the concept of putting more than one monolithic piece of silicon into a package to make a full system and package device. There's a BCM2710A1 SoC (system on a chip) in there and half a gigabyte of DRAM. These are both gold-wire stitched to the substrate. That is then encapsulated to make the package. And so, by doing this, we can get two chips into the space of one.

"Because the chips are similar sizes you need to use a silicon interposer, or spacer, amongst the centre of the processor. Then, when you put the memory on top, it doesn't squash the gold wires that connect the processor to the substrate. The memory is very sensitive, so there's less desire to put anything on top of the memory, which is why the SoC goes on the bottom and the memory on top.

"The A53 was the highest performance chip that we've ever used that used wire bonding in

" Because the chips are similar sizes you need to use a silicon interposer "

Know the numbers

Get to know what the numbers on the Raspberry Pi Zero 2 W's SiP mean:

RP30A0-AU: This is the SiP design number. The A0 refers to Raspberry Pi Zero, and the AU indicates the use of gold wiring. The first models used copper wires but it was changed to more expensive gold wiring to maximise long-term reliability).

2041: The week of manufacture (read right to left). Ours indicates the 41'st week in 2020.

200826: The date the batch was ordered (read right to left). In this case, 26 August 2020.

the package, and Raspberry Pi Zero 2 W displays a considerable uptick in performance. Single-threaded performance is approximately 40% faster than a normal Raspberry Pi Zero. Obviously, this is quite cool, but if you do a benchmark of the single-core on a Raspberry Pi Zero and compare it to the quad-core of Raspberry Pi Zero 2 W, you get more than five times the performance. Real time performance is around three times the speed. A Raspberry Pi Zero takes around 90 seconds to boot into the GUI, whereas Raspberry Pi Zero 2 W takes around 30 seconds.

"It's a small board, and if you are willing to help out by putting a heatsink on it, or if you put it in some sort of metal case where the heat can be drawn away from the chip, then it is possible to add voltage to the device so you can run it faster." A Raspberry Pi Zero 2 W with a cooling solution can generally sustain 1.2GHz performance."

Eben: "We use thick copper inside the board. So, effectively, we dissipate the heat throughout the board. So if you look at Raspberry Pi Zero 2 W with a thermal camera, you'll see that it gets hot throughout the board. That's because we're using copper to move heat away from the CPU.

"The can [Wireless Lan enclosure] is a forward-looking thing to the days where people want to design this into things. We've got to be upfront: there's not a huge amount of stock for the first year because 'hey, there's a global semiconductor shortage.' Looking ahead, this is a product that's going to be around for a long time, and we hope people will put it into product designs for OEMs."

Raspberry Pi Zero 2 W QuickStart Guide

Setting up and using Raspberry Pi Zero 2 W is pretty straightforward

Congratulations on becoming a Raspberry Pi Zero 2 W owner. We're sure you'll enjoy discovering a whole new world of computing and the chance to handcraft games, control robots, build machines, and share your experiences with other Raspberry Pi fans.

Getting started won't take long: just corral the extra bits and bobs you need on our checklist. To get set up, use Raspberry Pi Imager to set up a card and connect all the cables. This guide will lead you through each step. You'll find Raspberry Pi OS, including coding programs and office software, all available to use. After that, the world of digital making with Raspberry Pi awaits you.

What you need

All the bits and bobs you need to set up a Raspberry Pi computer

8GB microSD card

You'll need a microSD card with a capacity of 8GB or greater. Raspberry Pi Zero 2 W uses it to store the operating system and store programs and files. If you want to reuse an old card, you'll need a card reader: either USB or a microSD to full-sized SD (pictured).

Linux, Mac, or Windows computer

You'll need a Linux PC (such as another Raspberry Pi), Windows Linux PC, or Apple Mac computer, to run Raspberry Pi Imager to download and write Raspberry Pi OS onto a microSD card for Raspberry Pi Zero 2 W.

Power supply

Raspberry Pi Zero 2 W uses the same type of micro USB power connection as many old electronic devices. So you can recycle an old USB to micro USB cable and a smartphone power supply. Raspberry Pi also sells official power supplies (**magpi.cc/products**), which provide a reliable source of power.

USB keyboard

Like any computer, you need the means to type commands, create code and documents, and otherwise control Raspberry Pi. You can use a Bluetooth keyboard, but the initial setup process is much easier with a wired keyboard. Raspberry Pi sells an official Keyboard and Hub. **magpi.cc/keyboard**

Display and HDMI cable

A standard PC monitor is ideal, as the screen will be large enough to read comfortably. It needs to have an HDMI connection, as that's what's fitted on your Raspberry Pi board. Raspberry Pi Zero 2 W needs a mini HDMI to HDMI cable (or adaptor). **magpi.cc/minihdmi**

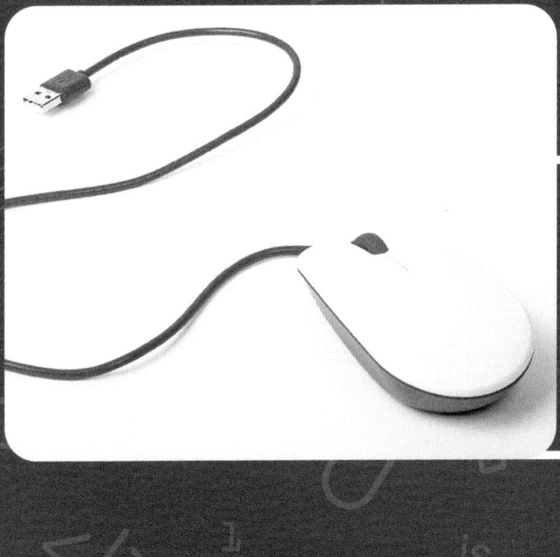

USB mouse

A tethered mouse that physically attaches to your Raspberry Pi via a USB port is simplest and, unlike a Bluetooth version, is less likely to get lost just when you need it. Like the keyboard, we think it's best to perform the setup with a wired mouse. Raspberry Pi sells an Official Mouse (**magpi.cc/mouse**).

Set up Raspberry Pi Zero 2 W

Set up your microSD card and connect all the accessories before powering up

A HDMI cable, such as one used by most modern televisions, is used to connect Zero 2 to a TV or display. You'll need a mini-HDMI to HDMI cable (or adaptor)

A wired mouse or keyboard is connected to the micro USB socket. Most devices use a USB-A connection, and you may need a micro USB to USB-A adaptor (**magpi.cc/usbadaptor**).

01 Get it connected

As you're setting up a Raspberry Pi Zero 2 W, you'll need to use a USB-A to micro USB-B cable (or adaptor) to connect a keyboard to the smaller connection on a Raspberry Pi Zero W. One is included with the official keyboard, which also has additional USB-A ports to accept the mouse and other accessories. Or you can use a cable with your own keyboard.

02 Mouse and keyboard

You can either connect your mouse to a USB socket on your keyboard (if one is available), then connect the keyboard to the micro USB socket (via the micro USB-A to USB-B adapter). Connect your mouse to one of these if possible. If not, you'll need a USB HUB so you can connect both the keyboard and mouse to the single USB-B socket.

03 More connections

Now connect your full-sized HDMI cable to the mini-HDMI to HDMI adapter, and plug the adapter into the mini-HDMI port in the middle of

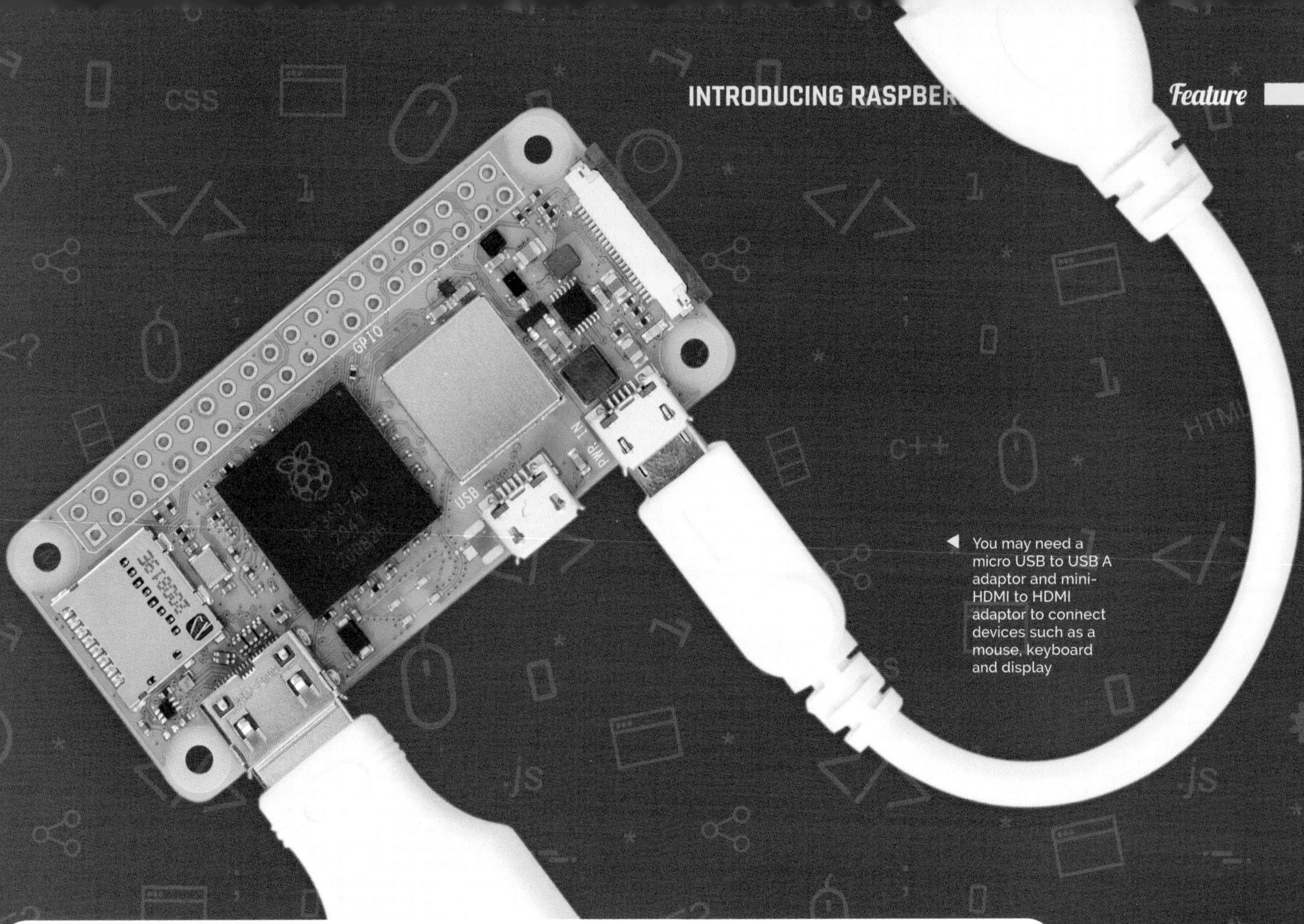

You may need a micro USB to USB A adaptor and mini-HDMI to HDMI adaptor to connect devices such as a mouse, keyboard and display

your Raspberry Pi Zero 2. Connect the other end of the HDMI cable to an HDMI monitor or television.

04 Set up the software

Now you've got all the pieces together, it's time to install an operating system on your Raspberry Pi so you can start using it. Download Raspberry Pi Imager from **magpi.cc/imager**. This utility is available for Windows, macOS, and Linux computers, so choose the relevant version for your system.

05 Write the OS to the microSD card

Attach your microSD card to your PC or Mac computer, and launch Raspberry Pi Imager. Click the 'Choose OS' button to select which operating system you would like to install. The top option is Raspberry Pi OS (32-bit). With an OS selected, click the 'Choose SD card' button and select your microSD card (typically there will be just one option).

Finally, click the 'Write' button and wait while the utility writes the selected OS to your card and then verifies it. When complete, you may remove the microSD card.

> " When Raspberry Pi OS first loads, you will need to set a few preferences "

06 Assemble your Raspberry Pi

Now it's time to physically set up your Raspberry Pi. Plug your PC monitor into the mains. Remove the microSD card from the SD card adaptor and slot it into the underside of your Raspberry Pi Zero 2 W.

07 Power up

Plug in your Raspberry Pi power supply and, after a few seconds, the screen should come on. Raspberry Pi OS will boot up. When Raspberry Pi OS first loads, you will need to set a few preferences. Click Next, when prompted, then select your time zone and preferred language, and create a login password. You're now ready to get online. Choose your WiFi network and type any required password. Once connected, click Next to allow Raspberry Pi OS to check for any OS updates. When it's done so, it may ask to reboot so the updates can be applied. You're all set to start enjoying computing with your very own Raspberry Pi Zero 2 W. M

SUBSCRIBE TODAY FOR JUST £10

Get 3 issues + FREE Pico W

+

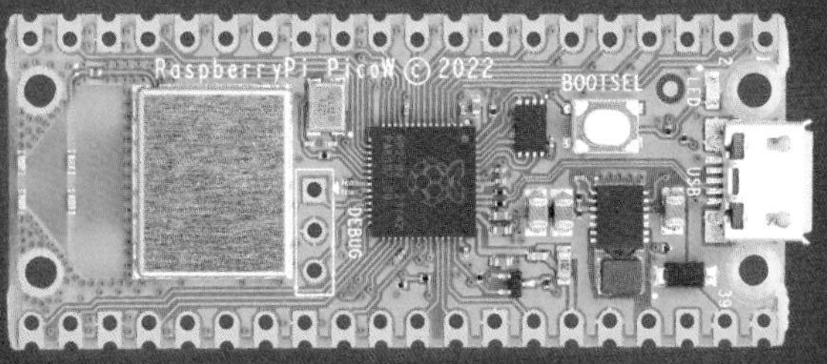

Subscriber Benefits

- **FREE Delivery**
  Get it fast and for FREE
- **Exclusive Offers**
  Great gifts, offers, and discounts
- **Great Savings**
  Save up to 35% compared to stores

Subscribe for £10

- Free Pico W
- 3 issues of The MagPi
- Free delivery to your door
- £10 (UK only)

Subscribe for 12 Months

- Free Pico W
- 12 issues of The MagPi
- Free delivery to your door

£55 (UK) £90 (USA)
£80 (EU) £90 (Rest of World)

Subscribe by phone: **01293 312193**

Subscribe online: **magpi.cc/subscribe**

Email: **magpi@subscriptionhelpline.co.uk**

Subscribe for £10 is a UK-only offer. The subscription will renew at £30 every six months unless cancelled. A free Pico W is included with a 12-month subscription in USA, Europe and Rest of World. (No Raspberry Pi Pico W with £5 Rolling Monthly Subscription).

SUBSCRIBE TODAY AND GET A

FREE Raspberry Pi Pico W

Subscribe in print today and get a **FREE development board**

WORTH **$6**

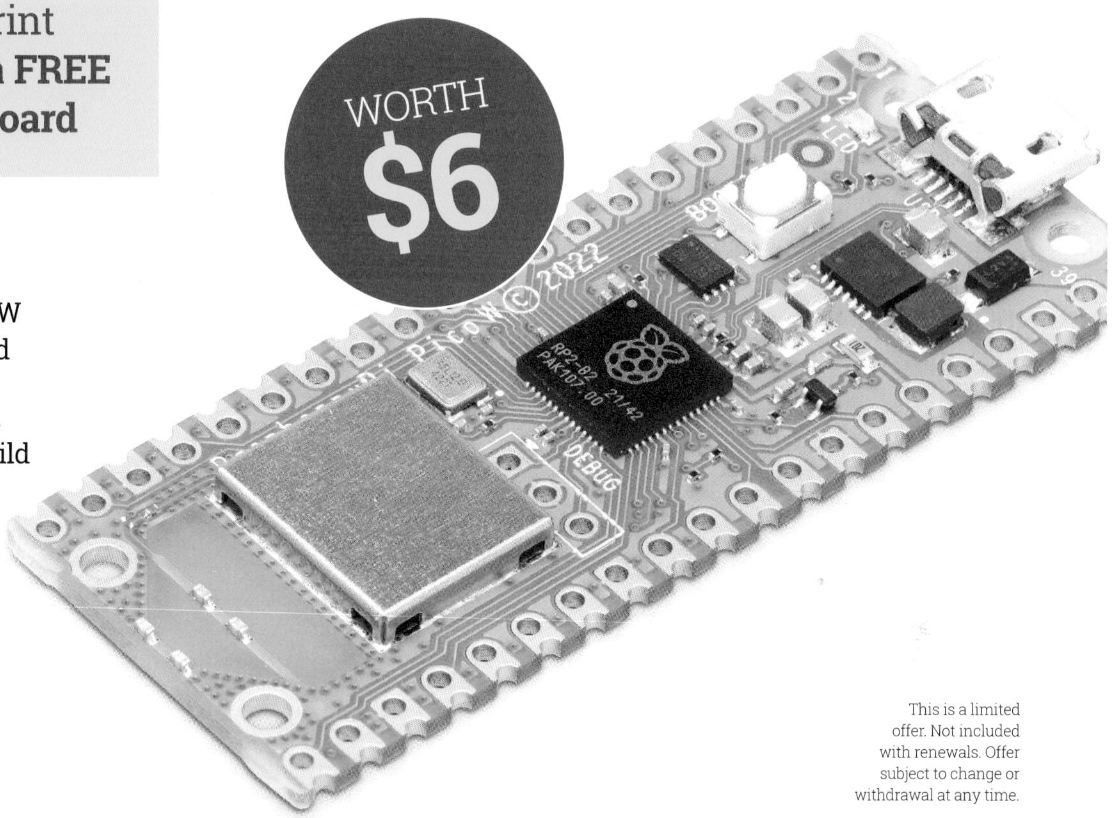

- A brand new RP2040-based Raspberry Pi Pico W development board
- Learn to code with electronics and build your own projects
- Make your own home automation projects, handheld consoles, tiny robots, and much, much more

This is a limited offer. Not included with renewals. Offer subject to change or withdrawal at any time.

Buy now: magpi.cc/subscribe

SUBSCRIBE on app stores

From **£2.29**

Available on the App Store

GET IT ON Google Play

Project Showcases

76

94

98

102
LIGHT
AUTO
ALARM
GOBLIN TEASMADE

▲ Art had to be rescaled, but it's been done faithfully

3/4 Star Wars Arcade Cabinet

Why pay over the odds when you can build an accurate replica, and have fun doing it? **Rob Zwetsloot** switches off his targeting computer to have a look

MAKER

James Milroy

A postman who loves making and fixing things in his free time, and spent his childhood in arcades.

@james_milroy

Getting the arcade machine of your dreams gets a little harder every day, especially the older they are. Making one, however, is always possible if you have the right skills and a Raspberry Pi.

"My project was to build a replica, or as close as I could reasonably manage, of the Atari Star Wars arcade cabinet," James Milroy tells us. "I really wanted to build a cockpit as that's what I played on in the eighties, but sadly I didn't have the room to house it, so the compromise was to build a stand-up cabinet instead."

Even then, the standard cabinet has a lot of detail, and James really nailed the look of it. Why build it from scratch, though? "Initially, I had toyed with sourcing an original cabinet and restoring it, but soon gave up on that idea after finding it nigh on impossible to source a cabinet here in the UK," James explains. "Almost all cabinets for sale were located in the USA, so they were out of the question due to the high cost of shipping. Atari only made just over 12,500 cabinets worldwide, so their rarity meant that they commanded top dollar, effectively putting them out of my price range. It was at this point that I decided that if it was going to happen, then I would have to make it myself."

Making a cabinet is hard enough, but the control system would have to be an original Atari yoke. "The Atari yoke is considered the 'holy grail' of controllers and, again, is very hard to find," James says. "My prayers were answered in October 2018 when a thread on a forum I was subscribed to popped up with a small Utah-based startup aiming to supply replica yokes at a realistic price to the arcade community. I grabbed two of these (one for my friend) and the project was on."

Good feeling

When it came to actually emulating the game, for James there was only one choice: "My decision to go with a Raspberry Pi was a no-brainer really. I had previously made a bartop cabinet using a Raspberry Pi 3 and RetroPie/EmulationStation which I was really pleased with. So I had a

The cab is made with MDF, and includes the original art
The yoke is an accurate replica that connects to Raspberry Pi via USB
The riser was a custom build by James that emulates lights from the films

Quick **FACTS**

- The original game came out in 1983...
- The same year as *Return of the Jedi*
- It used special vector graphics to emulate 3D space
- A Picade X-HAT handles everything but the replica yoke
- A lot of custom work was done to downsize stuff to 3/4 scale

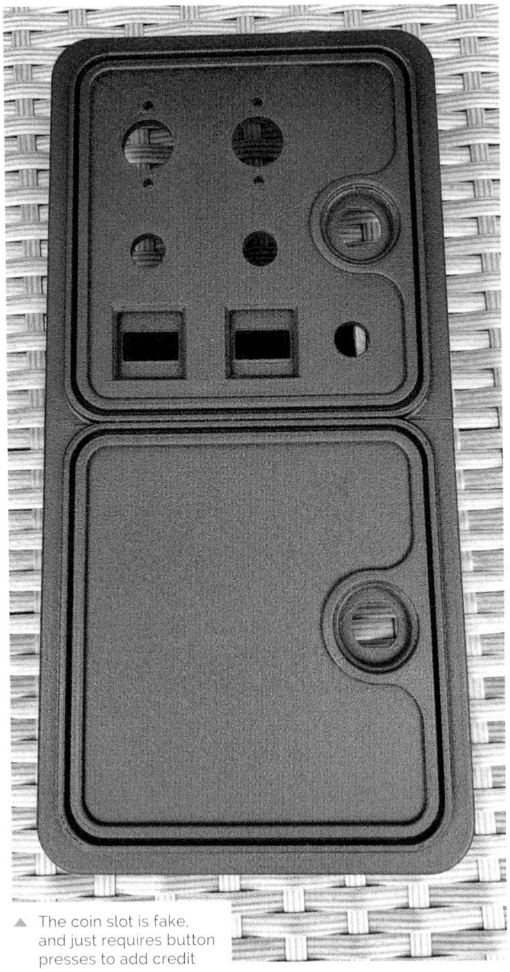

The coin slot is fake, and just requires button presses to add credit

3D-printing the mouldings and details is easier than recreating them with vacuum forming

> Overall, I'm really pleased with the way the cabinet has worked out

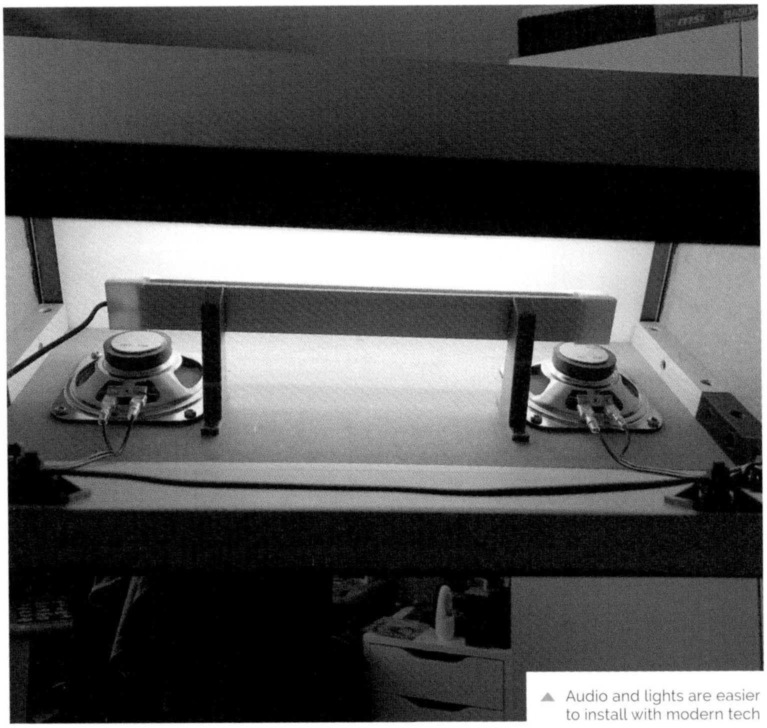

Audio and lights are easier to install with modern tech

platform that I already had experience with and knew was more than capable of emulating the one game I needed to run. Besides, the simplicity and low cost of the ecosystem for Raspberry Pi far outweighs the extra expense and effort required going down the PC route."

With a custom build and emulation, authenticity of the gameplay experience could be a bit off. However, that's not the case here. "I think that it plays just like the real arcade machine mainly due to the inclusion of the replica yoke controller, and adding your credit by pressing the button on the coin door," says James. "Ideally a vector monitor or a CRT would go a long way to making it look just like the original, but a reasonable representation is possible on an LCD using shaders and anti-aliasing. Gameplay does seem to get really hard really quick, though; this could be due to an imperfect emulation, but is more likely due to my reactions having dulled somewhat in the last 38 years!"

Always in motion

While the current build is amazing as it is, James does have some ideas to improve it. "Overall, I'm

▲ The wooden parts are laser-cut for added precision

really pleased with the way the cabinet has worked out," he says. "I will be replacing Raspberry Pi 3B+ with a Raspberry Pi 4 to enable me to run a newer version of MAME which will hopefully offer a better emulation, sort some audio glitching I get with my current setup, and hopefully enable some graphical effects (such as bloom and glow) to make it look more like its running on a CRT."

Accurately replicating

ATARI
STAR WARS
THE COIN VIDEO EXPERIENCE

01 "Luckily... I managed to have a play on an original cabinet at the Game On exhibition at Glasgow Braehead Arena. Armed with my trusty tape measure, I got some vital measurements from the real cabinet and after that, everything fell into place."

02 The cabinet was designed and laser-cut from MDF. There's some steel in the construction, as well as tempered glass over the screen. Mouldings are 3D-printed, with the help of a graphic artist who also helped rework graphics to fit the smaller size. Lastly, a riser is constructed to make sure it's tall enough to play.

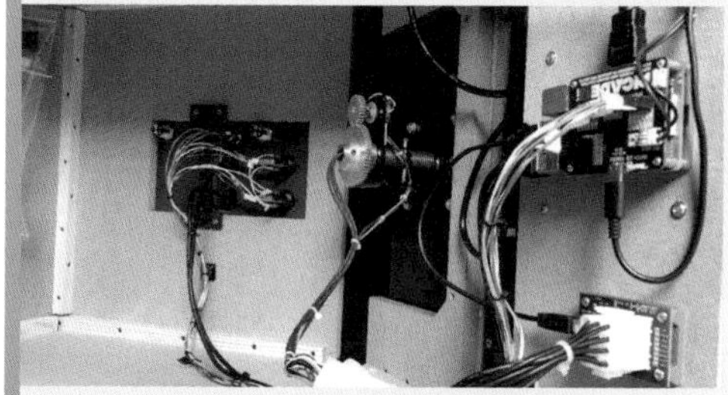

03 "The workings were simple when it came down to it: Raspberry Pi 3B+ with Pimoroni Picade X HAT. This gives us a power switch, audio amp, buttons, and a joystick if necessary. The replica yoke is interfaced with a USB adapter from the same company. It allows us to use the yoke with the original Atari connector. Software is RetroPie and I'm currently emulating on AdvMAME."

Epigone Drone

Inspired by NASA's attempt to launch a helicopter on Mars, one maker made an Earth-bound one of her own, hears **Rosie Hattersley**

MAKER

Avra Saslow

Avra studied computer science and geography in Boulder, Colorado, having got hooked on electronics and Raspberry Pi while making glass speakers for her final physics project.

instagram.com/avranator47

Like millions of us, in April Avra Saslow watched with bated breath as NASA's Perseverance rover touched down on the surface of Mars.

Like most of us, Avra knew all about the other ground-breaking feat being trialled alongside Perseverance: a helicopter launch called Ingenuity, that was to be the first flight on another planet – "a fairly lofty goal", says Avra, since "the atmosphere on Mars is 60 times less dense than Earth's."

With experience of Raspberry Pi-based creations, Avra was keen to emulate Ingenuity back here on earth (**magpi.cc/ingenuity**).

NASA chose to use open-source products and use commercially available parts for its helicopter build. It just so happened that Avra had recently begun working at SparkFun, a Colorado-based reseller that sells the very same Garmin LIDAR-Lite v3 laser altimeter that NASA's helicopter is based on. "It's a compact optical distance measurement sensor that gives the helicopter 'eyes' to see how far it hovers above ground," Avra explains.

A Garmin LIDAR-Lite v4 LED altimeter acts as 'eyes' and measures changes in the surface elevation

To avoid being swiped by the drone's rotors, the Raspberry Pi 4, which uses NASA's especially written F Prime code for telemetry, had to be positioned very carefully

The Epigone piggybacks on a borrowed DJI drone and is able to fly successfully for several minutes, despite a load equal to its own weight

Quick FACTS

- Avra is a big fan of Python for citizen science projects
- As a keen cyclist, Epigone's mapping plays to her love of wild terrains
- Epigone means 'a less distinguished follower or imitator'
- Avra loves Perseverance/ Ingenuity for its diversity of makers
- Ingenuity's lead engineer helped Avra complete her own version

▲ NASA recognises that Raspberry Pi offers a way to "dip your toe in embedded systems," says Avra, and "encourages the idea that Linux can run on two planets in the solar system"

" NASA posted the Ingenuity helicopter's open-source autonomous space-flight software on GitHub. It was written specifically for use with Raspberry Pi! "

▲ NASA's Ingenuity helicopter, which inspired the Epigone, completes its first one-way flight on Mars. (Image credit: NASA)

Alert!
Drone safety

Be mindful of spinning blades when modifying drones. Drone usage in the UK is regulated by the Civil Aviation Authority. Make sure you read the The Drone and Model Aircraft Code before flying your drone (and research relevant drone regulations in other countries).

magpi.cc/dronecode

NASA posted the Ingenuity helicopter's open-source autonomous space-flight software, written specifically for use with Raspberry Pi, on GitHub. Avra took all this as a sign she "just had to experiment with the same technology they sent to Mars."

F Prime and shine

Her plan was to see whether she could get GPS and lidar working within NASA's framework, "and then take the sensors up on a drone and see how it all performed in the air." Helpfully, NASA's GitHub post included a detailed F Prime tutorial based around Raspberry Pi: **magpi.cc/fprimegit**. Avra says understanding and using F Prime (F´) was the hardest part of her Epigone drone project. "It's a beast to take on from an electronics enthusiast standpoint," she says. Even so, she emphatically encourages others to explore F´ and the opportunity to make use of NASA's code: **magpi.cc/fprime**.

The Epigone Drone is built around Raspberry Pi 4 Model B; Garmin's LIDAR-Lite v4, which connects to a Qwiic breakout board and has a laser rather than an LED; a battery pack; and a DJI Mini 2 drone borrowed from a videographer colleague. Having seen how small the drone was, Avra realised 3D-printing an enclosure case would make everything far too heavy. As it was, positioning the Epigone onto its host drone was challenging enough: the drone's rotors passed worryingly close to the project's Raspberry Pi, even when precisely positioned in the centre of the drone's back. The drone has its own sensors to allow for controlled navigation, which meant Avra's design had to diverge from NASA's and have its lidar 'eyes' on its side rather than underneath.

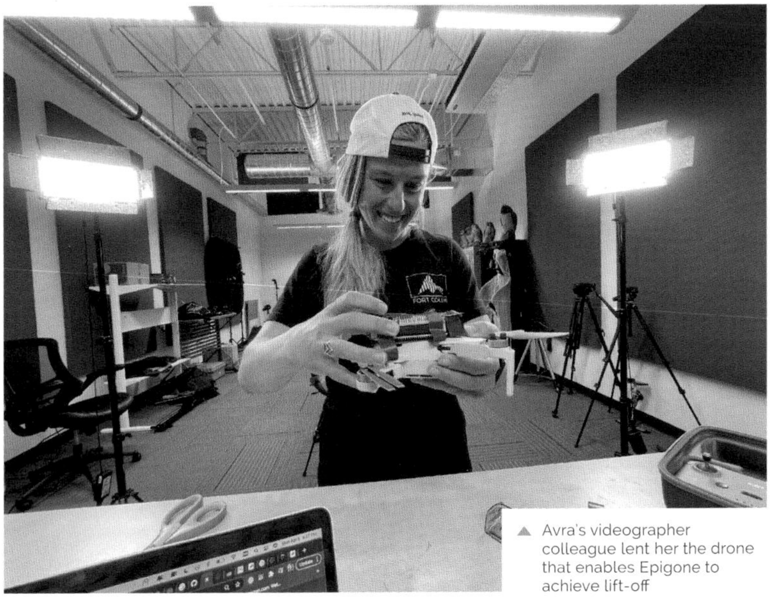

▲ Avra's videographer colleague lent her the drone that enables Epigone to achieve lift-off

▲ These glass speakers, made for her physics final project, were Avra's first foray into Raspberry Pi and Arduino

Although her version piggybacks on an existing drone, Avra was amazed when her Epigone creation took flight. "I honestly thought [it] would be too heavy to achieve lift, but what do ya know, it flew! It went up maybe 30 ft and we were able to check the sensors by moving it close and far from the SparkFun HQ [where she works]."

While the drone's battery depleted in "a matter of minutes" due to its additional load, the Epigone worked well and could be deployed to map small areas of land such as elevation changes in a garden, Avra suggests.

Make a Mars helicopter

01 To emulate Avra's version, you need Raspberry Pi 4, a Garmin LIDAR-Lite v4 for telemetry, an NMEA GPS receiver, Qwiic breakout board, a power pack, and a drone to provide the lift. You should closely follow NASA's detailed F´ tutorial for Raspberry Pi (**magpi.cc/fprimegit**). The steps assume you have your own drone to power your craft.

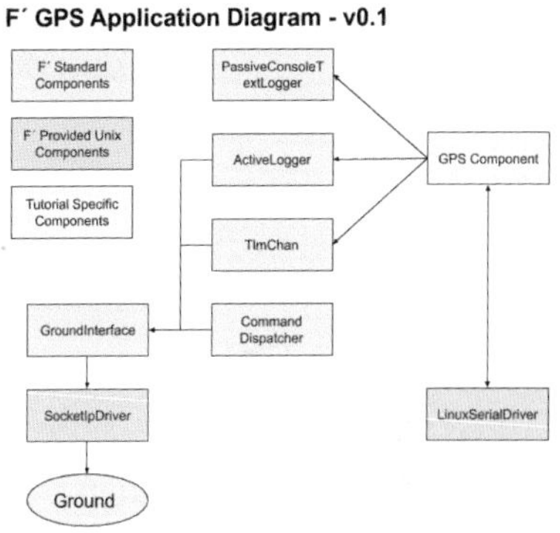

02 NASA explains how to use F´ with Raspberry Pi and an NMEA-enabled GPS receiver attached to a serial driver (**magpi.cc/fprimegps**). It can read in GPS messages from a UART port, then produce events and telemetry through the GPS link.

```
<?xml version="1.0" encoding="UTF-8"?>
<?oxygen RNGSchema="file:../xml/ISF_Component_Schema.rnc
<!-- GPS Tutorial: GpsApp/Gps/Commands.xml

This defines a single command to report the lock status
when working with GPS to determine if the data should be
-->
<commands>
    <!-- Define a single command that runs asynchronousl
    the GPS component's command space.  The mnemonic is
    <command kind="async" opcode="0" mnemonic="Gps_Repor
        <comment>A command to force an EVR reporting loc
    </command>
</commands>
```

03 Avra built a lidar component using NASA's GPS template to create the component through XML, then deployed the component on her native host and cross-compiled it for Raspberry Pi.

Bluebot Shoal Fish Robot

If you loved the film Finding Dory, you might just enjoy the original story of these underwater robots, suggests **Rosie Hattersley**

MAKER

Florian Berlinger

Florian is fascinated by displays of collective intelligence and co-operation such as schools of fish and teams of robots.

florianberlinger.ch

It's no coincidence that the shoal of robot fish in this Raspberry Pi Zero W project look more than a little like Dory from Pixar's movie. As with the film character, the Bluebot robot fish are based on the blue tang or surgeonfish. Unlike Dory, however, these robot fish are designed to be anything but loners. They behave collectively, which is the focus of the Blueswarm research project that began in 2016 at Harvard University.

Florian Berlinger and his PhD research project colleagues Radhika Nagpal, Melvin Gauci, Jeff Dusek, and Paula Wulko set out to investigate the behaviour of a synchronised group of underwater robots and how groups of such robot fish are co-ordinated by observing each other's movements. In the wild, birds, fish, and some animals co-ordinate in this way when migrating, looking for food and as a means of detecting and collectively avoiding predators. Simulations of such swarm behaviour exist, but Blueswarm has the additional challenge of operating underwater. Raspberry Pi Zero W works well here because multiple Bluebot robots can be accessed remotely over a secure wireless connection, and Raspberry Pi Zero W is physically small and light enough to fit inside a palm-sized robot.

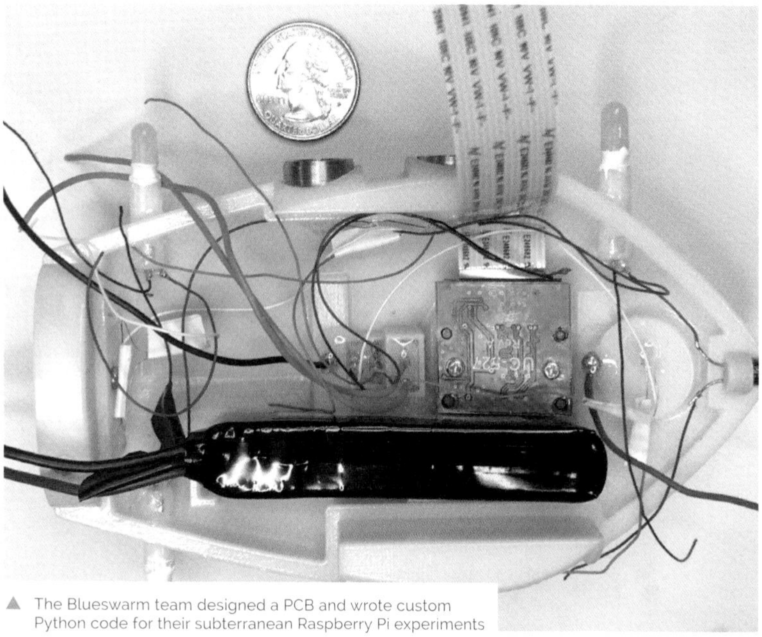

▲ The Blueswarm team designed a PCB and wrote custom Python code for their subterranean Raspberry Pi experiments

Mimicking movements

The team designed the fish-inspired, 3D-printed robot body as well as the fin-like actuators and the on-board printed circuit board which connects to all the electronics and communicates with Raspberry Pi Zero W. Designing the robot fish took the team four years, from working out how each robot fish would move and adding sensing capabilities, to refining the design and implementing collective behaviours, coded using Python 3.

They used as many off-the-shelf electronics as possible to keep the robots simple, but adapted existing software algorithms for the purposes of their investigations, "with several clever twists on existing algorithms to make them run fast on Raspberry Pi," adds Florian.

On-board cameras that offer "an amazing 360-degree field of view" are one of the project's real triumphs. These cameras are connected to Raspberry Pi via a duplexer board (so two cameras can operate as one) the project team co-designed with Arducam (see **arducam.com**). Each Raspberry Pi Zero W inside follows the camera images and instructs the fins to move accordingly. The team developed custom algorithms for synchronisation, flocking, milling, and search behaviours to simulate how real fish move individually and as a group. As a result, says Florian, "Blueswarm

Linked cameras attached to Raspberry Pi Zero W monitor what surrounding fish are doing. The Bluebot robot then mimics their behaviour, such as moving its fins
Once assembly of the electronics is complete, the 3D case is permanently sealed and all instructions issued wirelessly over a VPN
Illumination from the three LEDs on each fish robot makes image processing more efficient, since low light levels are harder to process

Quick FACTS

- There's a video about the Bluebot shoal on YouTube (**magpi.cc/swarmoffish**)
- The Bluebots are already being used for other studies on marine behaviour
- Blueswarm are going to study open water marine behaviour next
- When tackling complex projects, break them into smaller tasks…
- Start simple, develop and test each stage individually, Florian advises

> " On-board cameras offer an amazing 360-degree field of view "

can be used to study inter-robot co-ordination in the laboratory and to learn more about collective intelligence in nature." He suggests other robot-based projects could make use of a similar setup.

Imitation of life

Each robot fish cost around $250 and took approximately six hours to make. To make your own, you'd need a 3D printer, Raspberry Pi Zero W, a soldering station – and a suitably large tank for your robot shoal! Although the team hasn't made the code available, the Blueswarm project paper has recently been published in Science Robotics and by the IEEE Robots and Automation Society (**ieee-ras.org**). Several biology researchers have also been using the Bluebot shoal as 'fish surrogates' in their studies of swimming and schooling.

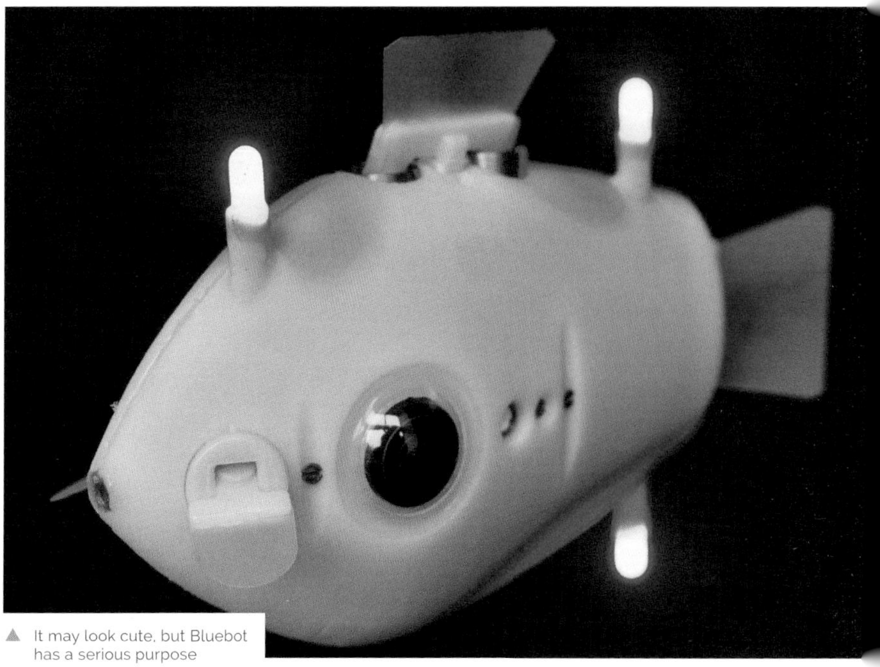

▲ It may look cute, but Bluebot has a serious purpose

Callisto II

A retro gaming fan gave his Raspberry Pi a 1980s makeover. **Rosie Hattersley** is on the case

MAKER

Kevin Solar

Game developer and software engineer Kevin loves designing physical builds and often uses the powerful capabilities of Raspberry Pi in his projects.

magpi.cc/solarcomputers

In his spare time, games developer and maker Kevin loves to tinker with all things retro: "I like to design new NES games using 6502 Assembly" he tells us, deftly setting the scene for how his Callisto II retro computer design came about. A hardware-based project, Callisto marries Kevin's enjoyment of 3D design and printing and his abiding love of retro gaming.

"It started in 2019 when I really wanted to 3D-print a full-sized retro or a terminal style of computer," says Kevin. "When I was looking all over the web, I was surprised that this sort of thing didn't exist. I saw lots of mini 3D printable retro computers and they were really good, but I wanted a full-size one that I could use for everyday tasks. Since this didn't exist, I had to make one."

Raspberry Pi was the obvious choice for the hardware to power his dream of recreating the look and feel of a 1980s computer. "I needed a desktop OS to make my retro computer very functional," he explains, "[and] Raspberry Pi is an inexpensive and very capable computer."

Kevin has previously designed two retro computer cases. The first, Callisto J-29, "was very rough around the edges" while the second one, Europa, "was too heavily inspired by the original Macintosh." For Callisto II he wanted to top these first designs and "make something that was very easy to print and assemble, but still looked great." He wasn't prepared to compromise on computing power either - hence his choice of Raspberry Pi.

Tricky curves

Despite his experience with 3D printing to date, Kevin says the trickiest part of the design was modelling the curves, as he'd not done this before. He persevered, knowing the curves would set his design apart from others. It was also a challenge to find a true 4:3 LCD screen that was inexpensive and readily available, he says, but rarer still to find a retro computer that used the 16:9 aspect ratio of modern displays. He was eventually able to source an 8-inch Pimoroni display to give Callisto that all-important 1980s look.

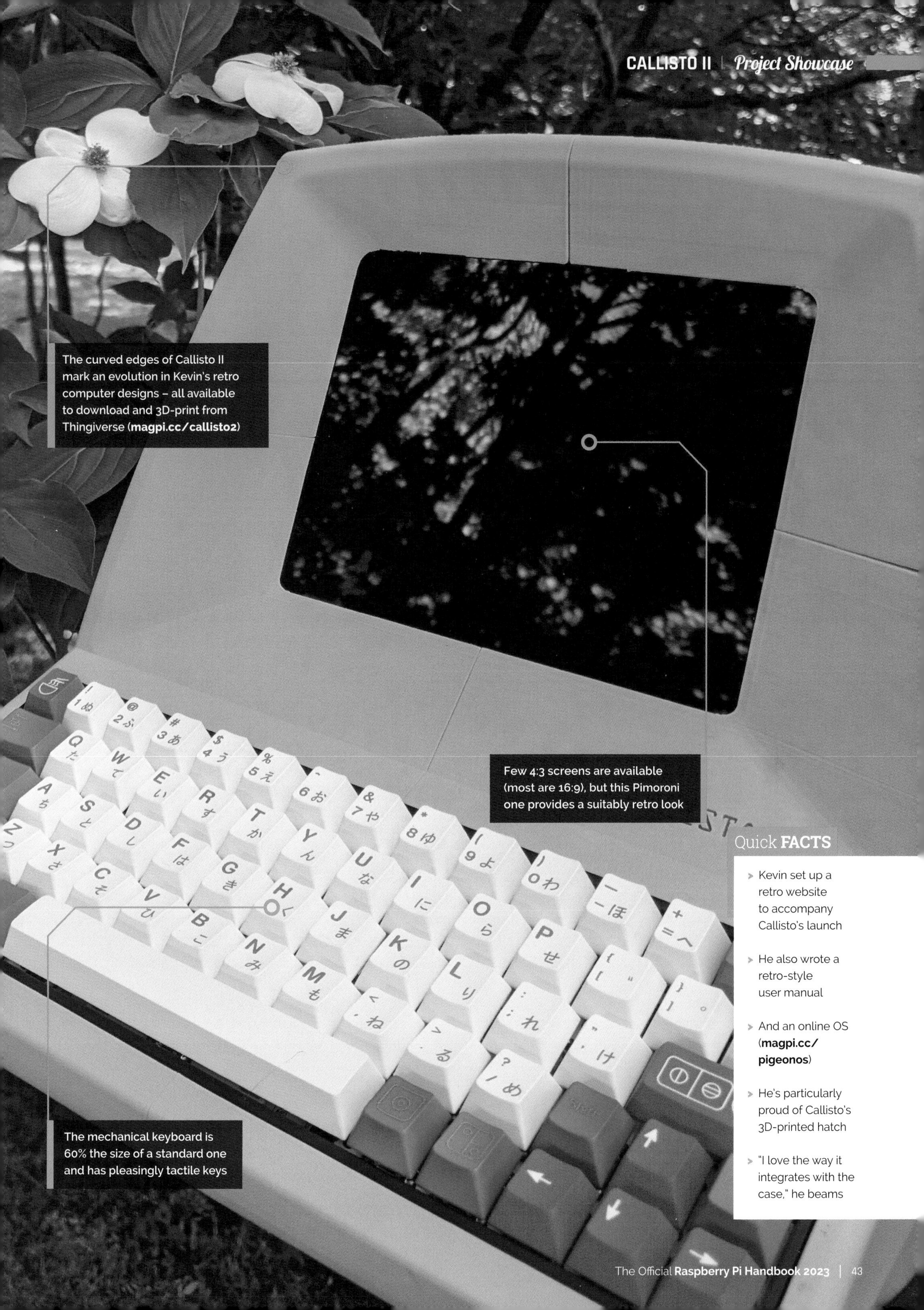

The curved edges of Callisto II mark an evolution in Kevin's retro computer designs – all available to download and 3D-print from Thingiverse (**magpi.cc/callisto2**)

Few 4:3 screens are available (most are 16:9), but this Pimoroni one provides a suitably retro look

The mechanical keyboard is 60% the size of a standard one and has pleasingly tactile keys

Quick FACTS

- Kevin set up a retro website to accompany Callisto's launch
- He also wrote a retro-style user manual
- And an online OS (**magpi.cc/pigeonos**)
- He's particularly proud of Callisto's 3D-printed hatch
- "I love the way it integrates with the case," he beams

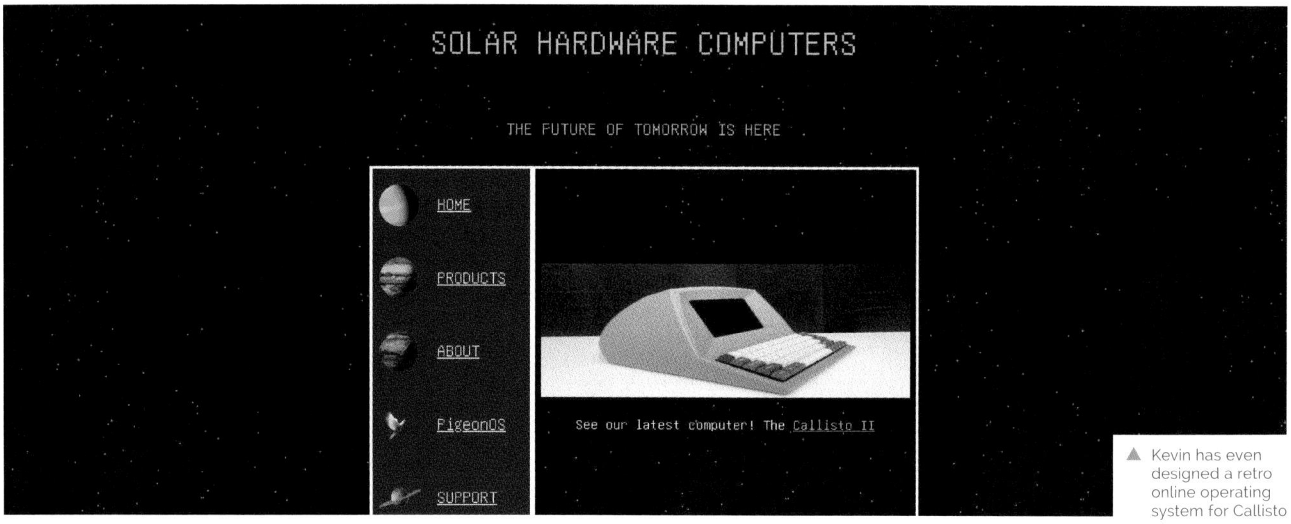

▲ Kevin has even designed a retro online operating system for Callisto

▲ Callisto II is the third 3D-printed retro computer Kevin has designed

Callisto's design is inspired by several real retro computers such as the ADM-3A, Zenith Z-89, and Hazeltine 1500, but with no 3D-printable, full-size retro computers available online for comparison, Kevin was on his own when it came to working out the dimensions and 3D design.

Easy ethos

Kevin tried to use readily available parts from online retailers for most of the project. "Not only did I want this to be super-easy to print and put together, I wanted it to be easy to find the parts [and be] something you could put together for an easy weekend project (assuming you spent the previous week printing all the parts)," he says. You don't even need glue as all the parts have been designed to snap together, but you should ensure all the electronics have room to breathe, he cautions.

▲ A trio of Callisto II computers

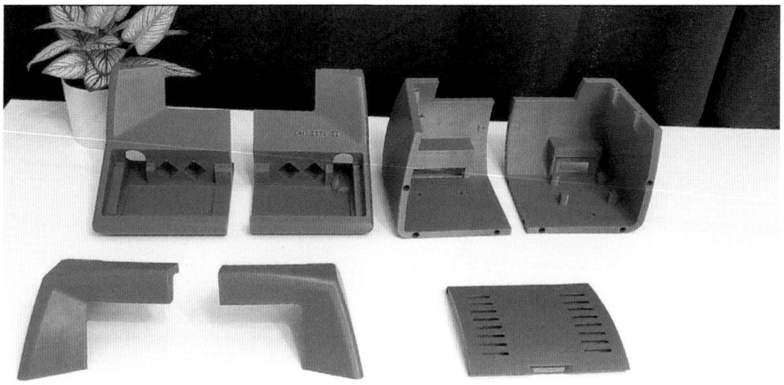

▲ 3D-printed parts ready for assembly

> Something you could put together for an easy weekend project

The project cost roughly $250 and involved printing six parts on a Prusa Mini 3D printer that each took a day to print. When sourcing a suitably tactile 60% mechanical keyboard, Kevin suggests choosing one that has blue switches. "These give the loudest clicks," he says.

Kevin stuck with Raspberry Pi OS, but part of Raspberry Pi's appeal is that you can load games and emulators to make it look and run however you want, he says. RetroPie is an obvious choice here. For more ideas on mimicking Callisto's retro looks with retro programs, take a look at Retro Computing in *The MagPi* issue 88 (**magpi.cc/88**).

▼ As Callisto II is 3D-printed, you can have it any colour

Build a retro computer

Download and 3D-print the parts for your Callisto II from **magpi.cc/callisto2**. The case is designed to snap together with pins; glue is optional.

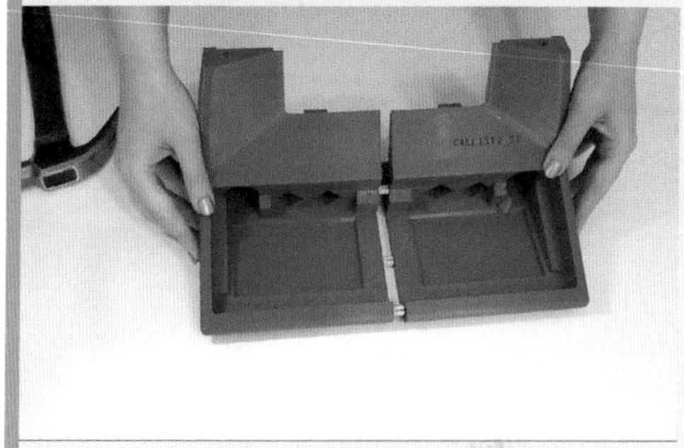

01 Assemble the case and install the hardware. The screen should easily slide into place using the case's built-in slots.

02 Slot the keyboard in place, followed by the USB hub and power supply. Secure them using 3D-printed pins.

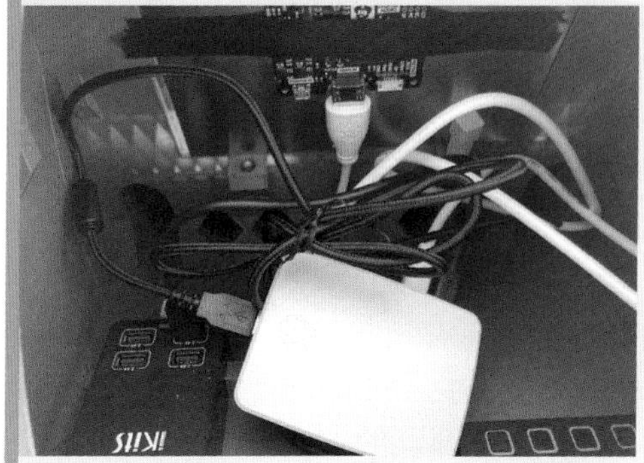

03 Insert and plug in Raspberry Pi near the PSU and, optionally, secure it using Velcro, then power up your retro computer.

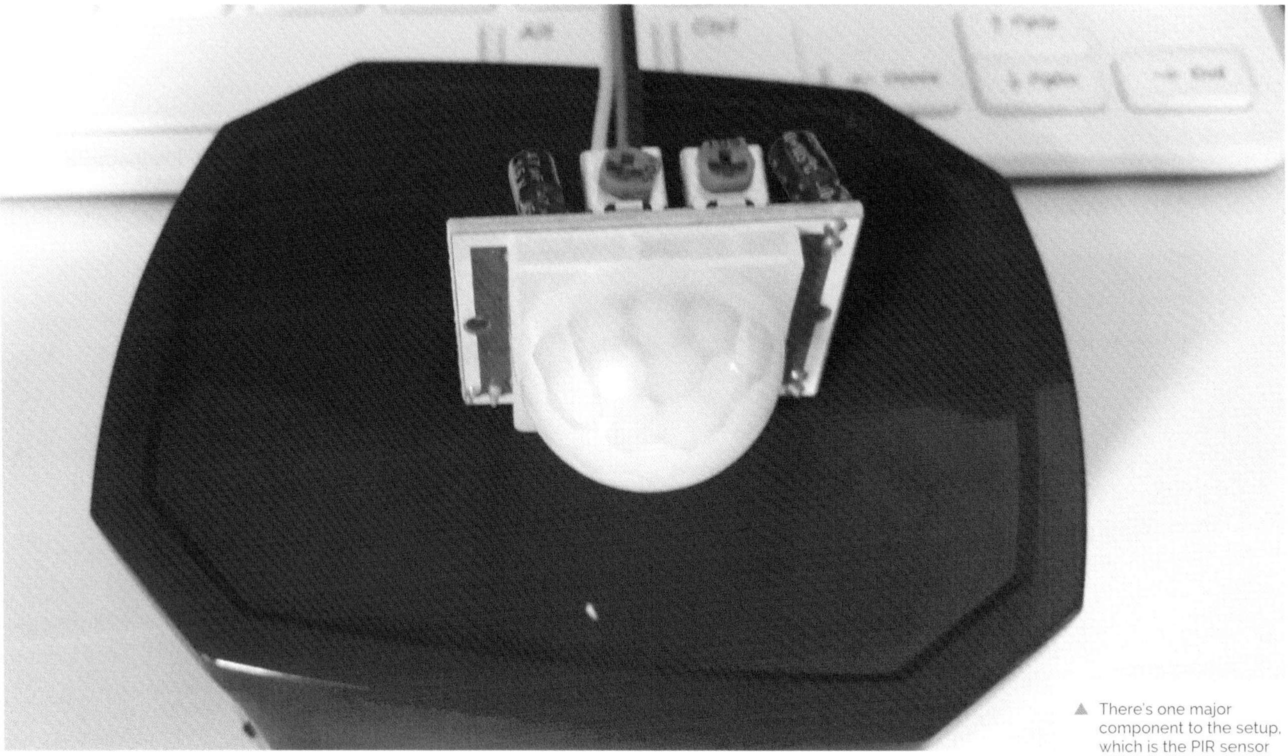

▲ There's one major component to the setup, which is the PIR sensor

Humane mousetrap

Safely catching mice is a better way of fixing a problem, and using Raspberry Pi means it needs less supervision. **Rob Zwetsloot** takes a look

MAKER

Andrew Taylor

A web developer who inherited a Raspberry Pi, allowing him to reconnect with the basics of computing that got him interested in the first place.

ataylor.net

With some IoT projects, it's the little things that help. For example, take Andrew Taylor, who did the good thing of setting up a humane mousetrap. However, checking it to see if any mice had been caught in it, while necessary, was getting a little boring.

"If a mouse had gone in and I did not check it, the mouse would quickly run out of food and water!" Andrew tells us. "Having been interested in Raspberry Pi for a couple of years and having recently begun learning Python using the Enviro+ environment sensors, I figured a Raspberry Pi with a motion sensor would be an ideal way to check."

It's a fairly simple setup, one commonly used in CCTV builds and some fun 'parent detectors' on the Raspberry Pi Foundation's projects site.

Mouse motion

"I came across a couple of automated mousetraps that people had made from scratch, but wanting to keep it simple and cheap," Andrew explains. "I

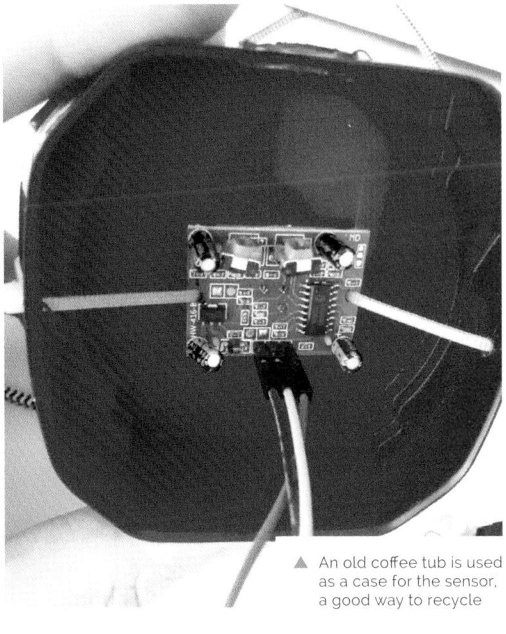

▲ An old coffee tub is used as a case for the sensor, a good way to recycle

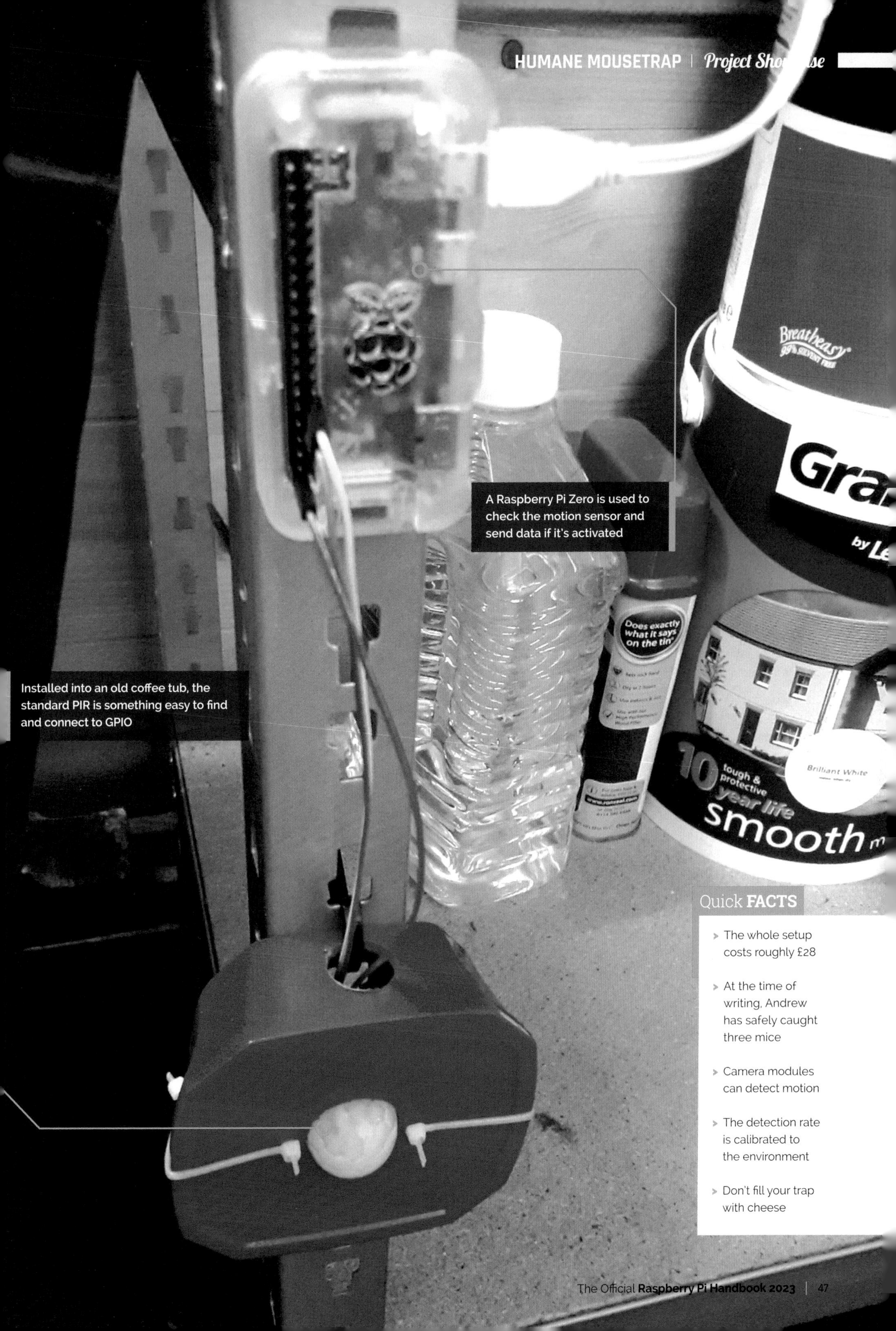

A Raspberry Pi Zero is used to check the motion sensor and send data if it's activated

Installed into an old coffee tub, the standard PIR is something easy to find and connect to GPIO

Quick FACTS

- The whole setup costs roughly £28
- At the time of writing, Andrew has safely caught three mice
- Camera modules can detect motion
- The detection rate is calibrated to the environment
- Don't fill your trap with cheese

▲ Wiring up the PIR to Raspberry Pi is quite simple, and means the project is easy to maintain

> I came across a couple of automated mousetraps that people had made from scratch, but wanted to keep it simple and cheap

wanted to use off-the-shelf parts where possible and keep costs down. The Pi Hut had a tutorial for a DIY burglar alarm utilising a PIR sensor, IFTTT, and Pushbullet, which seemed like an ideal starting point (**magpi.cc/pihutifttt**)."

IFTTT – If This Then That – is an online service popular with IoT folks. It's great for small things like cross-posting images on social media services, or sending a push notification when motion is detected in a mousetrap.

"I have only had one mouse since, but it worked!" Andrew says. "I was averaging about 800 detections a day and suddenly got well over a 1000. Sure enough, there was a mouse in the trap which I released shortly afterwards. I do tend to notice that the values fluctuate a bit, so it is always worth checking over the previous day's results to see if it is notably higher."

You might think that 800 push notifications a day is far worse than just occasionally checking your garage, and you'd be right, so Andrew tweaked the code a bit: "The code examples I found sent a

▲ It's a simple design, and was kept simple to keep to a small budget

▲ The first successful capture was released back outside the garage

notification for each movement detection – which I knew would be rather annoying, considering how randomly PIR sensors sometimes seem to trigger. My script instead logs any hits at a max of 1 per 30 seconds and then triggers a notification once every 24 hours, meaning I just get one notification a day."

Beat a path

There's always room for improvement, as Andrew explains: "I intend to improve the code so that it can record running averages and give an indication as to whether it believes there has been a significant spike that might necessitate me checking it out."

Whilst the aim of the project was to keep costs down, Andrew is tempted to experiment by adding a camera, and possibly a light, so he can have a peek remotely when there has been a spike in the readings and to see if it is a false alarm. Which, as he admits, is "a new height in laziness!"

Catching mice

01 The setup stays on 24/7, and monitors the PIR sensor for movement. The PIR uses infrared so it works just fine even in the dark.

02 Movement is tracked using a Python script which checks, and logs, every 30 seconds. This allows for more accurate long-term readings.

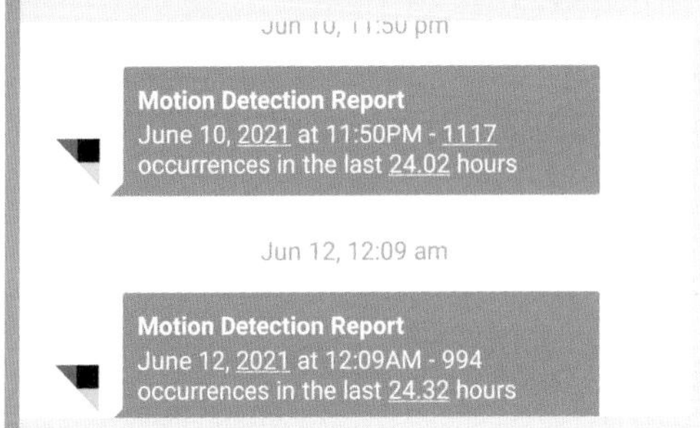
Motion Detection Report
June 10, 2021 at 11:50PM - 1117 occurrences in the last 24.02 hours
Jun 12, 12:09 am
Motion Detection Report
June 12, 2021 at 12:09AM - 994 occurrences in the last 24.32 hours

03 "After 24 hours, it triggers a message containing the number of hits over that period by sending a HTTP request to IFTTT which is hooked up to Pushbullet," says Andrew. "This then sends this message as a notification to my phone. The counter is then reset ready for the next day."

Campervan LAN

Needing to travel to several countries for work, one maker chose to kit out his campervan as a mobile office. **Rosie Hattersley** was intrigued

From the driving seat, Enrico can get live updates about on-board essentials such as water and fuel, as well as alerts about intruders

MAKER

Enrico Miglino

Enrico designs and develops multi-platform software projects for clients across Europe (and occasionally India). He's been using and helping road-test Raspberry Pi since its first Linux prototype.

we-are-borg.com

Campervans are all the rage right now, since they offer a chance to escape the home in favour of fresh air and a change of scene. Enrico Miglino's job often needs him to pitch up in Germany, Spain, or Belgium where his software development architect consultancy is in demand, so it made sense for him to take the digital nomad concept literally, and adapt a campervan to create a mobile office.

Although most campsites and caravan parks offer WiFi to guests, such connections are often unsecured, limited to only a couple of hours' free use, and require each device to log in separately. Enrico needed a setup that was far more robust. His vehicle, Jan The Van, sports a secure mobile LAN, is powered by three Raspberry Pi computers, and allows Enrico to use a single login to provide internet access to any devices on his network. The whole setup cost less than 500 euros, meaning he could also afford to add a solar panel.

Hack The Van

Enrico bought his van specifically for use as a mobile office. "The project aims to create a modular set of technological improvements in a standard vehicle to convert it to a secured and efficient mobile unit for living, working, and travelling," he explains.

It's named after Enrico's friend Jan Cumps who helped him work out the electronics that would be needed to successfully hack the van. The pair spent many evenings working together studying how to develop projects, debug hardware circuits, and teaching workshops at the Ingegno Makerspace in Ghent in Belgium (**ingegno.be**).

Jan The Van doubles as a mobile office and panel van conversion

Two Raspberry Pi 4 boards control the on-board LAN and provide wireless logins for passenger's smartphones, laptops, and tablets

Quick **FACTS**

- Enrico has an original Raspberry Pi prototype
- He participated in the design of its first Linux distro
- He loves Raspberry Pi for its flexibility and adaptability
- A modular approach means project components are readily reusable
- The biggest Jan The Van challenge: space, of course

▲ Jan The Van – a mobile office with a constantly changing landscape

▼ A quick glance inside and Jan seems like a relatively normal luxurious campervan

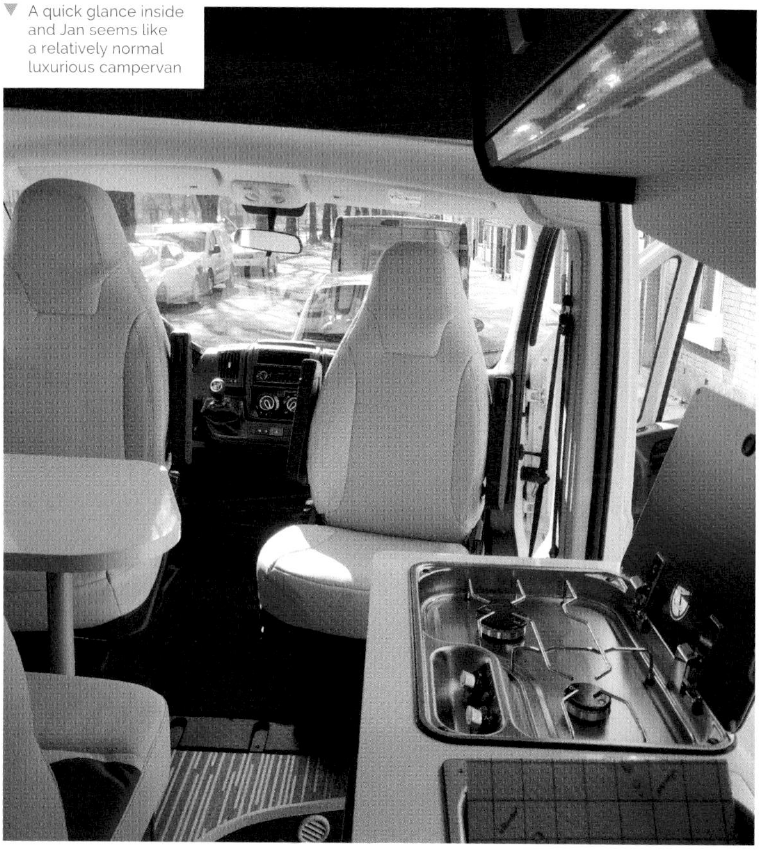

The duo took a modular approach to the build, starting with an 8GB Raspberry Pi 4B mounted in a case with a 7-inch touchscreen. Developed with Node-RED, the display shows the state of the network, firewall status, and the local weather. It also has screens monitoring any physical intrusions around Jan The Van.

DHCP features and the wireless LAN to Ethernet router make it safe for Enrico to connect his shielded network to insecure public WiFi networks. He's currently adding sensors to monitor gas and smoke levels from the van's kitchen, and to show how much water he's storing on board, providing reassurance there are plenty of provisions on lengthier, more remote journeys.

Enrico found lots of ideas online for how to create his vision, building and testing each element before installing it. "I tried to design it to be as modular as possible so it's easy to replicate it and adapt the modules to similar but not identical environments," he says.

More Raspberry Pi

Internal devices on Jan The Van's network could only be connected to the Ethernet port, which wouldn't work for iPads, iPhones, and other mobile devices that lack the necessary port. To overcome this, Enrico added an Ethernet switch and a second Raspberry Pi 4B configured as a bridge to which

▲ Jan The Van's third passenger reportedly enjoys the ride

> I tried to design it to be as modular as possible so it's easy to replicate it and adapt the modules

these mobile devices can connect wirelessly. A third Raspberry Pi connected to a full HD webcam provides live visuals of the rear of the vehicle while he's driving, and acts as a motion sensor security camera when the campervan is parked. It "detects motion around the camera field of view and records video just in case," Enrico says.

Jan The Van's first big trip was a ten-day journey from Belgium to a "nice campsite" in Spain – some 2800 km – where Enrico easily hooked up his campervan LAN and was able to work as a consultant each morning and tinker with his mobile office setup each evening. Planned improvements include adding a third Raspberry Pi 4 with a display to stream images from the van's rear camera and other information to the dashboard or cockpit. Most important, says Enrico, "is travelling to nice places to test the prototype in the real world."

Warning!
Electric hookup

Be careful when adding a power inverter to a van for electric hookup and seek professional advice.

magpi.cc/powerinverter

Hack your van

01 To create your own campervan network, you'll need two Raspberry Pi 4 computers for the LAN, one of which acts as a wireless bridge to provide ad hoc access, and a 300W or less power inverter. Laptops can also connect via Ethernet.

02 The main interface uses Node-RED software to manage the hardware, with C/C++ to program the ESP3266 modules attached to the sensors. The screen orientation is corrected in configuration.

03 Enrico used Python bash scripting to manage the network, execute timed tasks, and automate image saving from the on-board camera. This can be either a webcam or a Raspberry Pi Camera Module. Full details of Enrico's hack the van project are at **magpi.cc/janthevan**.

RFID Floppy Disk Reader

Ever wanted to make contactless floppy disks? Maybe not, but **Rob Zwetsloot** takes an appropriately distanced look at some

MAKER

Dylan Blake

A business analyst from Texas who codes in Python and SQL for a large grocery chain.

Old-looking tech with futuristic properties is a popular concept in media these days, so much so that we're surprised this is one of the first projects like this we've seen.

"My project was taking an old 1988 word processor and repurposing it into a gaming emulator with a Raspberry Pi," creator Dylan Blake tells us. "I wanted to utilise the floppy disk drive with RFID tags to initiate the software and have a working power button for added effect."

We've covered an RFID-based record player that did something similar with vinyls, but it wasn't built into an old piece of tech like this. "I came up with the idea by realising I didn't have a cool case to put my emulator in, and I really dig all things retro," Dylan explains. "I found this device for $20 on a marketplace app and thought it would be awesome to work with."

Breathing new life

The way the system works makes it feel almost like a classic computer, albeit a bit faster.

"When you click the tactile power button, you are briefly greeted with a retro splash screen and then cute computer ASCII art prompting to insert a floppy disk," Dylan says. "You fumble around for your favourite game handwritten on a 3.5-inch floppy, insert it into the floppy disk bay, and your game immediately starts up. If you don't know what game you want to play, you can insert the 'All' floppy to access the RetroPie game menu (of course, using the 8-bit theme)."

Dylan chose Raspberry Pi to power this for all the familiar reasons – a good size, easy access to GPIO pins, and it also allowed him to get more comfortable with Linux. "Raspberry Pi has fascinated me for years, and I probably have five of them at this point for various projects."

> "I especially like that my two-year-old son enjoys playing with it, albeit poorly"

Old-school cool

We're big proponents of learning things when building projects, and as well as getting more experience with Linux, Dylan learnt how to use RFID tags and readers in the process, which we think is a cool skill. It wasn't his first choice, though.

"I have a USB floppy reader that I would like to utilise instead of the RFID tag reader for look and feel," he admits. "But right now I like the ease of use of the RFID reader."

The original power button works to turn on the entire system
The real floppy disks have RFID tags, and the disk drive has an RFID reader that recognises what disk has been inserted
This classic word processor was upcycled by stripping nearly everything out

Quick FACTS

- As the printer has been removed, there's a lot of empty space on top
- Unfortunately, the original keyboard could not be used just yet
- This is because the keyboard connects via RJ45
- Word processors used to be machines, not just the name of PC software
- It uses an 800×600 display instead of the original

▲ Dylan's two-year-old likes to make use of it

He says the reactions he's received from it completely validate why he did it: "It was very popular on Reddit where I originally shared it, and my family loves how niche and fun it is! I especially like that my two-year-old son enjoys playing with it, albeit poorly. The only negative feedback I've received is that I repurposed a device that was already working. My counter to that is the word processor was only good for typing documents and saving them to floppies. Repurposing it has given life to this old tech, even if it's just the aesthetics of the original device."

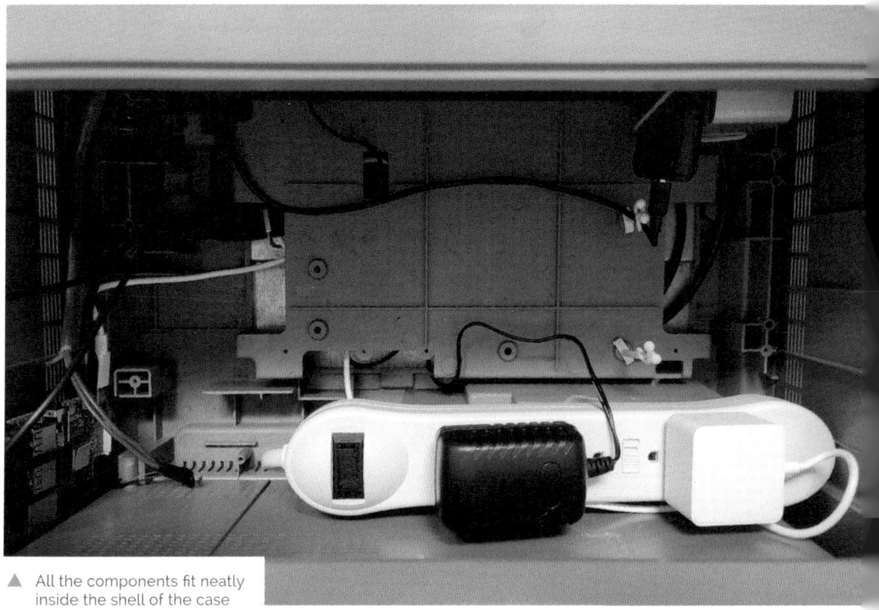

▲ All the components fit neatly inside the shell of the case

HIIT Workout Trainer

If watching the Olympics has motivated you to improve your fitness, here's a fabulous workout idea using a Raspberry Pi. **Nicola King** feels the burn

MAKER

James Wong

James is a machine learning engineer working on building industrial recommender systems and he enjoys building things with data science.

magpi.cc/hiitpi

When James Wong felt that his workout routine needed a boost due to extended periods of having to work from home, he knew he needed to take action. "First of all, working from home all day long without going to the gym has taken a toll on my health," he explains. "Secondly, as a machine learning practitioner, doing research on edge devices like Raspberry Pi and microcontrollers has always been fascinating to me, in the sense that the computational efficiency continuously pushes the boundaries of what we can achieve on a $30 computer with ML."

With that in mind, James married his regular HIIT (high-intensity interval training) workout with Raspberry Pi, and the HIIT Raspberry Pi web app was born. It's a clever idea that uses machine learning on Raspberry Pi to keep tabs on your workout in real time and make sure you are getting optimum results.

> It gamifies workouts with the leaderboard dashboard, and makes doing exercise fun with your partner and friends

▲ The Coral USB Accelerator's Edge TPU processor enables Raspberry Pi to analyse images at 30 frames per second

No pain, no gain

James first had the idea in January 2020, and had a prototype ready around three weeks later, with improvements and tweaks then made based on feedback from family, friends, and internet reaction. He's taken the HIIT concept, where you have short bursts of very high-intensity activity alternated with rest or much lower intensity exercise, and arguably improved it.

"It gamifies workouts with the leaderboard dashboard, and makes doing exercise fun with your partner and friends," he says. What's more, the app is simplicity itself in that it basically does just two things. Firstly, it uses computer vision, and a Raspberry Pi Camera Module, to track movements and poses, and then it scores them based on a set of predefined standards.

To speed up the machine learning algorithms, James plugged a USB Coral Accelerator Edge TPU into his Raspberry Pi. This, he highlights, was an important addition: "An Edge TPU is required for it to work roughly at 30 fps. Without the accelerator, frame rates drop significantly and the user experience deteriorates."

Of course, as with all builds, some fine-tuning was required, as James shares. "Every single workout consists of many moves or a sequence of poses. HIIT Workout Trainer needs to make sure to track them at a fraction of a second under various external conditions. Also, the web app interface has gone through many evolutions to have an intuitive and simplistic version of high-performance."

James has made various improvements to the project, with most coming from faster and more efficient video processing by the ML model under the hood, as well as user interface redesigns on the front. "I'm definitely going to test more workouts

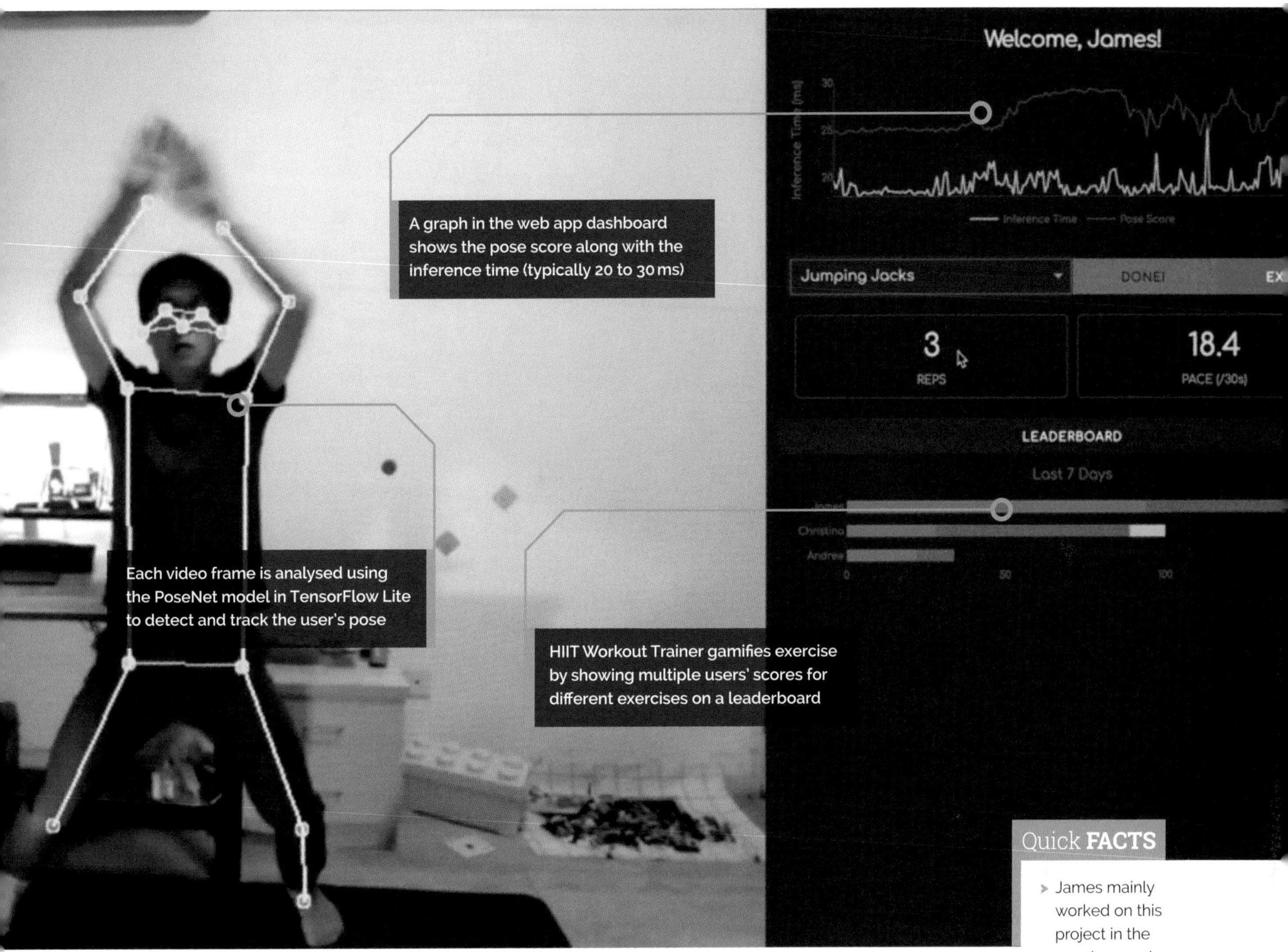
Welcome, James!
Inference Time (ms)
30
25
20
Inference Time
Pose Score
A graph in the web app dashboard shows the pose score along with the inference time (typically 20 to 30 ms)
Jumping Jacks
DONE!
3
REPS
18.4
PACE (/30s)
LEADERBOARD
Last 7 Days
Christina
Andrew
0
50
100
Each video frame is analysed using the PoseNet model in TensorFlow Lite to detect and track the user's pose
HIIT Workout Trainer gamifies exercise by showing multiple users' scores for different exercises on a leaderboard

Quick FACTS

- James mainly worked on this project in the evenings and at weekends
- He got some great feedback from YouTube and Reddit
- Find the code on GitHub: **magpi.cc/hiitpigit**
- This project by James can classify hundreds of birds and flowers: **magpi.cc/raspicamvision**
- James is working on a project to generate classical music on Raspberry Pi

if I have enough time on hand. It would be great if more people joined in," he says.

Unbiased adjudicator

This idea has obvious potential. James has described the app as akin to an "electronic referee", and it could easily be adapted for other forms of sport or exercise: "I've already seen people do similar awesome projects, but for weight training."

So, if you'd like to try your hand at this project, James very much encourages it. "The project is open-sourced on GitHub [**magpi.cc/hiitpigit**], where you can find more to set it up and get started. If anyone has any questions, please find me via email or Twitter [see **magpi.cc/jameswong**], I'm more than happy to help."

What more motivation do you need? As James reminds us on his web page, "Just imagining a workout is never the same as actually doing it. With everything put into place, let's slip on sweatpants and get it rolling!"

▲ A Camera Module provides the video frames which are analysed using machine learning on Raspberry Pi with the aid of an Edge TPU

▲ miniLIGO fires lasers at a beamsplitter to create an interferometer used for subatomic analysis

miniLIGO gravitational wave detector

A Nobel-prize winning concept inspired PhD students to make their own gravitational wave tracker using Raspberry Pi. **Rosie Hattersley** explains

MAKER

Husni Almoubayyed

Syrian-born PhD student Husni uses code and data to study the universe and often combines them with his Raspberry Pi maker projects

magpi.cc/miniligo

After hearing about the science prize-winning LIGO (Laser Interferometer Gravitational-wave Observatory, magpi.cc/ligo) six PhD students at Carnegie Mellon University in Pittsburgh decided to see whether they could emulate the idea.

LIGO detects the gravitational waves in space caused by celestial objects colliding and predicted by Albert Einstein in his General Theory of Relativity.

Understanding space has always interested Husni Almoubayyed, so he was attracted to the idea of something that can monitor gravitational effects "stretching and contracting space around us on a very tiny scale". While there was no expectation of building their own LIGO device, not least because of the expense involved, Husni and fellow students wanted to help communicate to a wide audience how LIGO works. miniLIGO is designed to be a "simplistic and affordable" prototype mimicking the laser-interferometer gravitational wave observatory experiment that won the Nobel Prize in physics in 2017.

Sensitive subject

The team had an advantage because Husni had worked on the original LIGO project when he was studying at Glasgow University as part of a scientific collaboration. "While I mostly worked on data analysis at that time, I wanted to experience some of LIGO's engineering challenges," he says, of which one is creating an interferometer (**magpi.cc/ligoifm**) for highly detailed analysis work. For example, CalTech's LIGO interferometers can measure a distance 1/10,000th the width of a proton!

Raspberry Pi 3B provides the processing power to analyse signals received and displays them on screen and, thanks to Python code, as sound

A beam splitter divides a laser beam into two, with mirrors used to bounce and merge them to create an interference pattern

A luminosity sensor analyses the 'fringes' of the object to discern minute details or phenomena, such as tiny distortions

▲ Team miniLIGO – Oshadha Gunasekara, Aileen Zhai, Husni Almoubayyed, Alisa Chang, Elliot Tuck, and Anirudh Mani – demonstrate their device at Build18, Carnegie Mellon University

Quick FACTS

- The $1,000 budget was set by Carnegie Mellon University
- miniLIGO was created for its Build18 event
- The first LIGO detector was launched two decades ago...
- The first gravitational wave was detected in 2015
- A space-based version of LIGO – LISA – is currently being planned

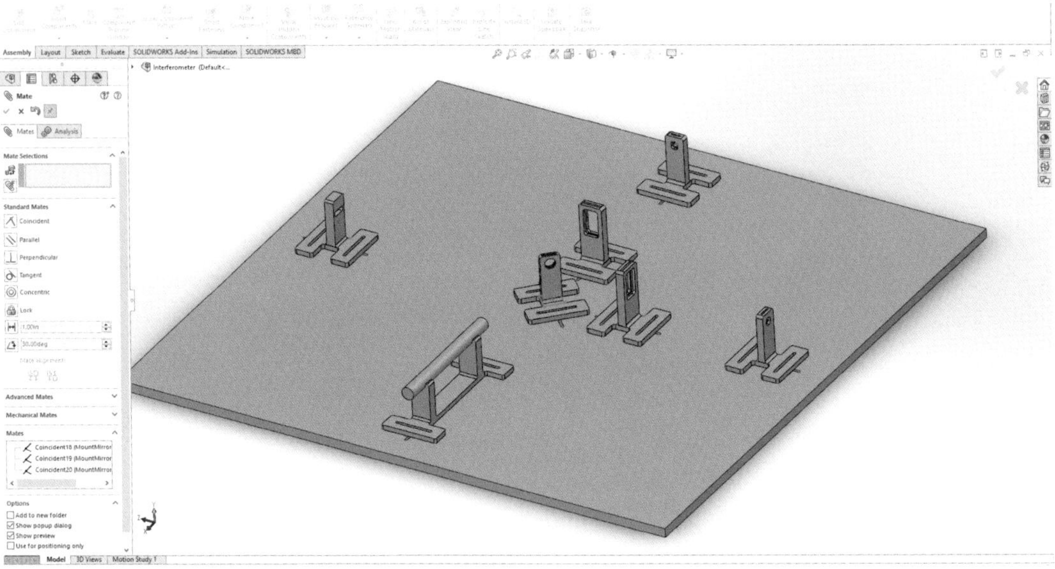

▶ Schematic of the interferometer used for miniLIGO

▼ The base is made of 2-way mirrored acrylic and laser-cut with annotations to indicate where the 3D-printed parts should fit

> LIGO has an extremely accurate damping system that works to shield it from vibrations and seismic changes

▼ Putting together miniLIGO ahead of its public demonstration

"LIGO is an extremely expensive and large-scale experiment," says Husni, and some major decisions were needed about how to create a prototype cost-effectively. With the aim of building the miniLIGO project for less than $1000, it was clear that some compromises would be needed, while mimicking the original LIGO concept as closely as possible.

"LIGO has an extremely accurate damping system that works to shield it from vibrations and seismic changes," Husni explains. "We considered using a six-axes motion sensor as a low-cost alternative, and to correct the signal computationally using the sensor, but ultimately decided to leave it out as we deemed that it would not be a good use of time and resources as stronger sources of noise would still dominate the signal."

Having used microcontrollers and Raspberry Pi for wearable computing and music-based projects, the possibilities of single-board computers were clear. (One of the other projects you'll find on the **husni.space** website is a graphical tablet that makes music with a corresponding artistic visualisation.) "Raspberry Pi allows us to easily connect all the pieces of the project, and has enough processing power to do signal processing in real time," he says. Raspberry Pi provides easy interaction with the light sensor, while also having plenty of processing power to analyse and process wave signals in real-time. It also meant the project could be portable – the team successfully ran miniLIGO from a battery pack.

Getting set up

A Raspberry Pi 3B, connected to a portable battery and a Bluetooth speaker, and a GPIO library are at the heart of miniLIGO. The team used an Adafruit High Dynamic Range Light Sensor and

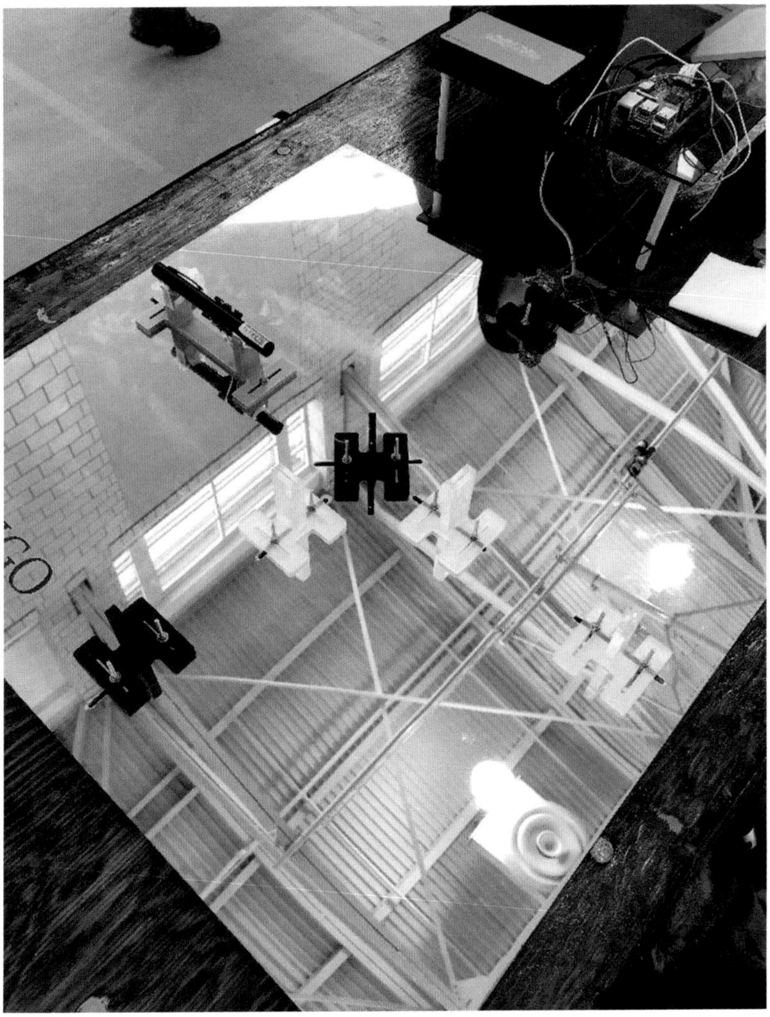

optical equipment, including the lenses and laser from Edmund Optics, then added a monitor and a speaker. Access to the university 3D printing lab meant they were able to 3D-print their own case.

When it came to writing software for miniLIGO, they decided to use Python code, largely because of the "mature and easy-to-use" Python library, from which they chose some of the common scientific computing libraries such as numpy, scipy, and matplotlib.

When miniLIGo's sensors detect a signal, its software analyses it, visualises it in real-time and plays a sound representation of it in Python. The team also wrote the software to simulate a gravitational wave signal whenever the red button on the device is pressed.

Husni explains: "miniLIGO is based on a Michelson interferometer setup that has several applications in physics and astronomy. Although it is nowhere near precise enough for gravitational wave detection, miniLIGO can be used to detect sources of noise that LIGO encounters, such as when lightning strikes nearby."

▲ The two perfectly reflective mirrors are used to build the Michelson interferometer

Warning!
Laser pointer

Be careful when using laser pointers in your projects and don't aim a laser pointer at a person's head.

magpi.cc/lasersafety

Make it mini

To make your own miniLIGO, you'll need a Raspberry Pi 3B powered by a portable battery pack. The code can be found at: **magpi.cc/miniligogit.**

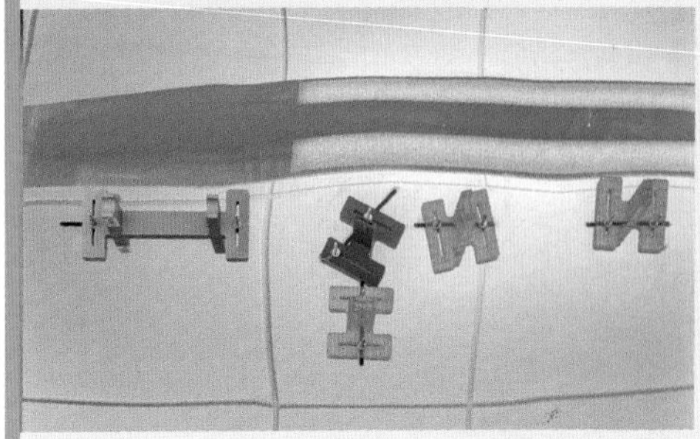

01 A high distance laser and a 50:50 beam splitter splits the laser into the two arms of an interferometer. You'll need two mirrors and two 75:25 beamsplitters to create two Fabry-Perot cavities.

02 Use an acrylic board as the base for the interferometer, and 3D-print or attach holders to it at exactly equal heights. Use an Adafruit High Dynamic Range Sensor or a Raspberry Pi HQ Camera to capture movements.

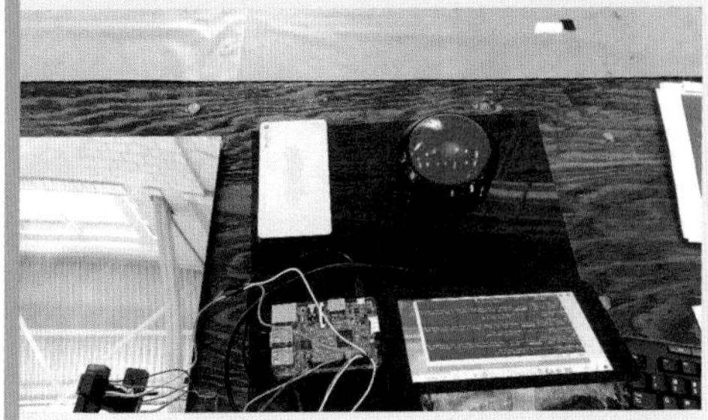

03 Attach a screen to show visualisations and a Bluetooth speaker to alert you to any waves detected. You could also add a trigger button to initiate demo visualisations.

Package Thief Deterrent

Here's a decidedly tongue-in-cheek make that is an extremely humorous take on a home security system. **Nicola King** stands well clear

MAKER

Ryder, aka 'Ryder Calm Down'

Ryder is a software engineer who likes to build ridiculous inventions in his spare time and document them on his YouTube channel Ryder Calm Down.

magpi.cc/ryderyt

Many of us are increasingly ordering goods online, with the consequence that more and more parcels and boxes are landing on doorsteps everywhere. From a security aspect, it can be a little worrying when our precious new purchases are left in plain sight waiting for us to return home. A fun and light-hearted project by Canadian maker Ryder shows one unique way of handling it.

Flour power

Following the theft of a package from his porch earlier in the year, Ryder decided to take action. "It had never happened to me before and I was a little frustrated," he recalls. "I'd been doing some image classification work at the time, and figured this would be a great opportunity to build something to deter people from stealing packages from my porch."

> It's great at detecting whether there is or is not a package, but it also sometimes recognises my cats as packages!

▲ A Raspberry Pi 4 runs the software, with a relay board used to control 12 V components such as the flashing light, siren, and sprinkler

Armed with a Raspberry Pi 4, a security camera, and various other ingredients, he set about making something that certainly caused a stir in the neighbourhood, as his must-watch YouTube video demonstrates (**magpi.cc/packagetheft**). "I built the project more for entertainment value than actual use (since I've only ever had one package stolen), but I consider it a success either way," he tells us.

So, how does the alarm system work? Using a security camera pointed at Ryder's door, images from the camera are pulled by his Raspberry Pi and processed by a custom machine learning model to detect if there is or isn't a package.

"If a package has been taken unexpectedly, Raspberry Pi sends signals to a relay (via its GPIO pins) to activate a variety of alarms to entice the 'thief' to drop the package," Ryder explains. Those surprises for thieves include a sprinkler, a loud truck horn, and a flour shower.

"This is definitely a great project if you're looking to get your feet wet into machine learning – though I don't recommend you implement the sprinkler, flour, or air horn components," he says. "All the participants in my video were friends and neighbours of mine, and implementing the exact same setup for the public will likely get you into some sort of trouble," he cautions. However, he does suggest that you can modify the project to send you an email when a package arrives, or has been taken unexpectedly, which may be a more prudent route to take!

Averting a cat-astrophe

Computer vision is central to Ryder's project, and he used Google Cloud AutoML to train the machine learning model. "I wanted something to include in my video that was easiest for those getting started with machine learning to use," he reveals. "Training a model yourself can be quite intimidating."

Warning!
Don't Assault

You may get into trouble if you attack innocent porch visitors with flour, or anything else for that matter. Do not deploy that aspect of the project in a real-world environment.

Quick FACTS

- This project took Ryder around a month to complete.
- You'll need quite a few items to make this...
- "A combination of Python, TensorFlow, and the last of my sanity!"
- Find the code on GitHub: **magpi.cc/packagetheftgit**
- He's working on a Raspberry Pi-powered autonomous car

It wasn't plain sailing to get this system working as he wanted. Ryder originally had all alarms and sprinklers go off at the same time, but decided to move them to separate threads with separate timings in order to best soak his very obliging friends and neighbours. "The model [is] great at detecting whether there is or is not a package," he says "but it also sometimes recognises my cats as packages!" He also sensibly trained the system to disarm when he arrived to pick up his parcels.

Many co-operative friends and neighbours have embraced his undertaking, "though I did buy my neighbours a few boxes of chocolate afterwards to say sorry for the noise", Ryder admits. For the time being, he has dismantled this project, "since I don't really have the need for it, but I'm brainstorming some version two ideas – one of which involves a net." Any future front-step felons may be in for a shock!

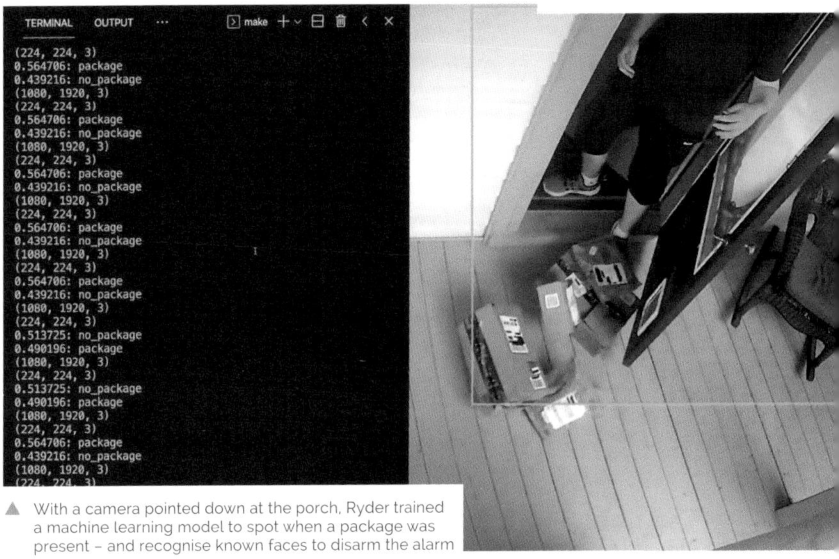

▲ With a camera pointed down at the porch, Ryder trained a machine learning model to spot when a package was present – and recognise known faces to disarm the alarm

▲ Wiring up the capture suit components

Machine Learning Prosthetic Arm

This intelligent arm learns how to move naturally, based on what the wearer is doing, as **Phil King** discovers

MAKER

James Bruton

Former toy designer, current YouTube maker, and general robotics, electrical and mechanical engineer, James is a fan of doing it yourself and innovation by trial and error.

magpi.cc/jamesbruton

Known for his robotic creations, popular YouTuber James Bruton is also a keen Iron Man cosplayer, and his latest invention would surely impress Tony Stark: an intelligent prosthetic arm that can move naturally and autonomously, depending on the wearer's body posture and limb movements.

"It's a project I've been thinking about for a while, but I've never actually attempted properly," James tells us. "I thought it would be good to have a work stream of something that could be useful."

Motion capture suit

To obtain the body movement data on which to base the arm's movements, James considered using a brain computer, but this would be unreliable without embedding electrodes in his head! So, he instead opted to train it with machine learning.

For this he created a motion capture suit from 3D-printed parts to gather all the data from his body motions: arms, legs, and head. The suit measures joint movements using rotating pieces with magnetic encoders, along with limb and head positions – via a special headband – using MPU-6050 inertial measurement units and Teensy LC boards.

Collected by a Teensy 4.1, this data is then fed into a machine learning model running on the suit's Raspberry Pi Zero using AOgmaNeo (**magpi.cc/aogmaneo**), a lightweight C++ software library designed to run on low-power devices such a microcontrollers.

"AOgmaNeo is a reinforcement machine learning system which learns what all of the data is doing in relation to itself," James explains. "This means that you can remove any piece of data and, after training, the software will do its best to replace the missing piece with a learned output. In my case, I'm removing the right arm and using the learned output to drive the prosthetic arm, but it could be any limb."

While James notes that AOgmaNeo is actually meant for reinforcement learning,"in this case we know what the output should be rather than it being unknown and learning through binary reinforcement."

To train the model, James used distinctive repeated motions, such as walking, so that the prosthetic arm would later be able to predict what it should do from incoming sensor data. He also spent some time standing still so that the arm would know what to do in that situation.

New model arm

With the machine learning model trained, Raspberry Pi Zero can be put into playback mode to control the backpack-mounted arm's movements

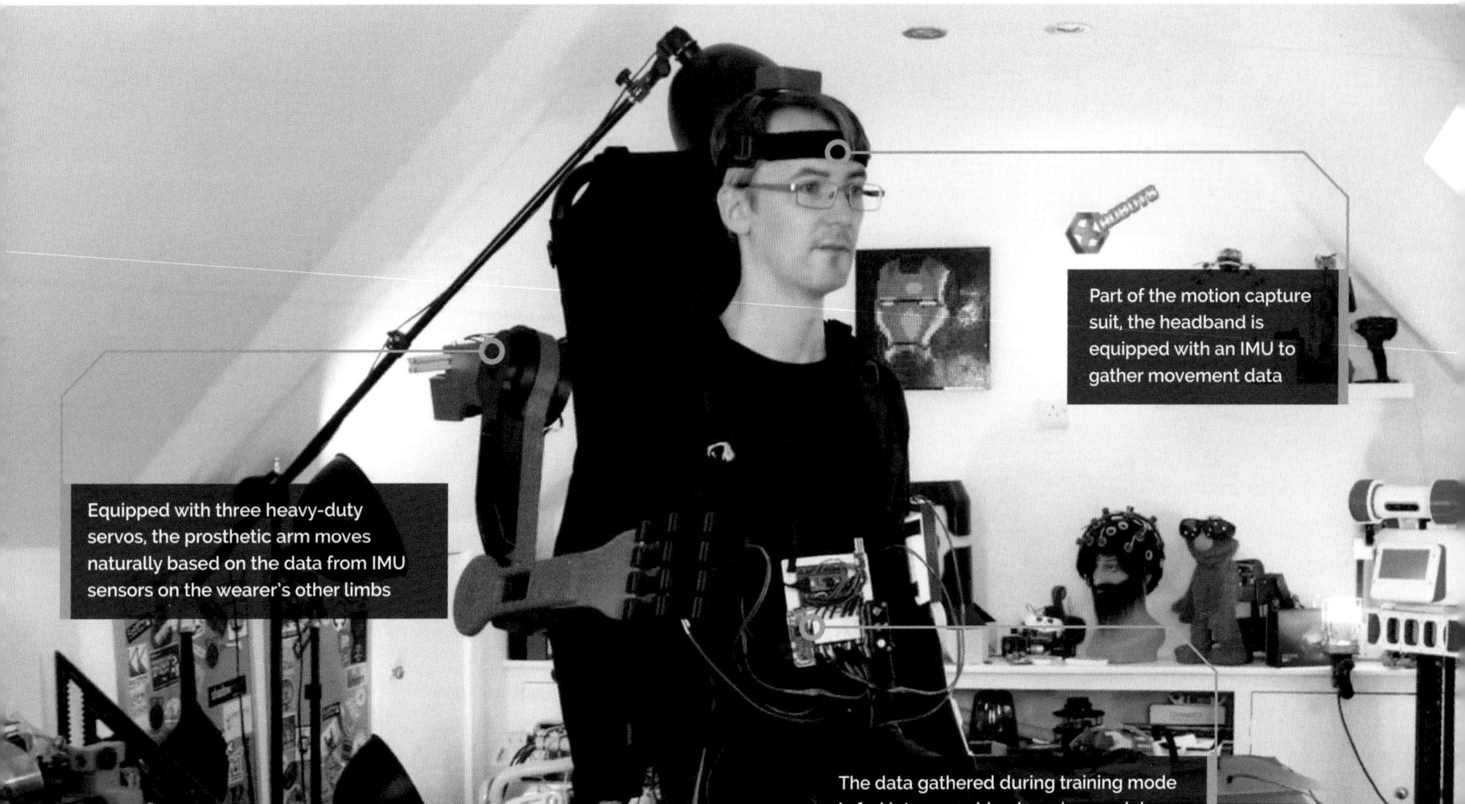
Part of the motion capture suit, the headband is equipped with an IMU to gather movement data
Equipped with three heavy-duty servos, the prosthetic arm moves naturally based on the data from IMU sensors on the wearer's other limbs
The data gathered during training mode is fed into a machine learning model on a Raspberry Pi Zero; in playback mode, it controls the robotic arm

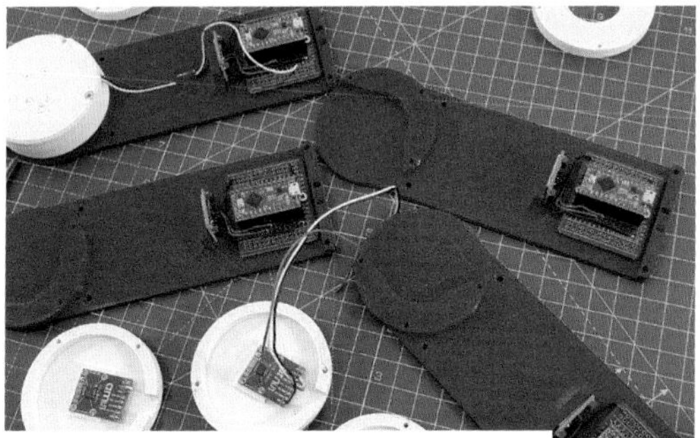

▲ The motion capture suit comprises 3D-printed parts, each equipped with a magnetic rotary encoder, MPU-6050 IMU, and Teensy LC

Quick FACTS

- The project took James a few months to create
- He designed and 3D-printed the robot arm parts
- It uses three high-torque XM540-W270-T servos
- The arm is mounted on 2020 aluminium extrusions
- These are fitted to a standard frame backpack

> Raspberry Pi Zero can be put into playback mode to control the backpack-mounted arm's movements intelligently

intelligently. It can then duplicate what the wearer's real right arm was doing during training depending on the positions and movements of other body parts.

So, as he demonstrates in his YouTube video (**magpi.cc/mlarmyt**), if James starts walking on the spot, the prosthetic arm swings the opposite way to his left arm as he strides along, and moves forward as raises his left leg. If he stands still, the arm will hang down by his side. The 3D-printed hand was added purely for aesthetic reasons and the fingers don't move.

James admits that the project is highly experimental and currently an early work in progress. "I'd like to develop this concept further," he says, "although the current setup is slightly overambitious and impractical. I think the next step will be to have a simpler set of inputs and outputs."

While he generally publishes his CAD designs and code, the arm "doesn't work all that well, so I haven't this time. AOgmaNeo is open-source, though (free for personal use), so you can make something similar if you wished." What would you do with an extra arm?

Oasis-grow

Growing fruit and vegetables can be so frustrating, but a Raspberry Pi-based smart agriculture project helps take the uncertainty out of cultivation. **Rosie Hattersley** reports

MAKER

Mike Lee and Aviel Stein

Oasis-X co-founders Mike and Aviel met at a maker event in Pennsylvania where self-taught coder and entrepreneur Mike was showing off several Raspberry Pi projects.

oasis-x.io

"Plants, mushrooms, and other cultivated organisms don't respond in linear ways," even when provided with supposedly 'ideal' growing conditions, much to the frustration and disappointment of those keen to cultivate their own food, observes 25-year-old entrepreneur and technology fan Mike Lee. Such nonconformity "is a mindset that many programmers and engineers find difficult to accept," says Mike, explaining why he and his PhD student co-founder Aviel Stein set about improving the odds of a successful harvest. Their response, Oasis-grow, is a 'smart agriculture engine' that runs on Raspberry Pi and collects detailed information about whatever the user cares to grow in a bid for more predictable results.

▶ The temperature-controlled incubation tent has sensors to monitor humidity, airflow, and light levels

The system provides visual and sensor data, plus controls for heat, humidity, airflow, light, and watering cycles. The selling point is a system that "keeps your plants, mushrooms, and Petri dishes happy, and remotely monitors the grow space so you can identify and address problems before they get serious."

Make room for mushrooms

Mike and Aviel met at the 2019 Pennsylvania Maker Faire where, among other projects of his own devising, Mike presented his first Raspberry Pi project: a dashboard camera with buttons and a remote, back-window LED screen for signalling other cars. "It was very cool but perhaps a bit too dangerous for the road (the other cars did not react ... well)," he admits. Nonetheless, the pair decided to work together, and quickly formed Oasis-X (**magpi.cc/oasisx**) having identified a market for a smart agriculture monitoring tool.

Aware that DIY plant monitoring and sensor kits were already available, Mike and Aviel needed to be sure they were offering something that fulfilled the scalability promise it would need to be of interest

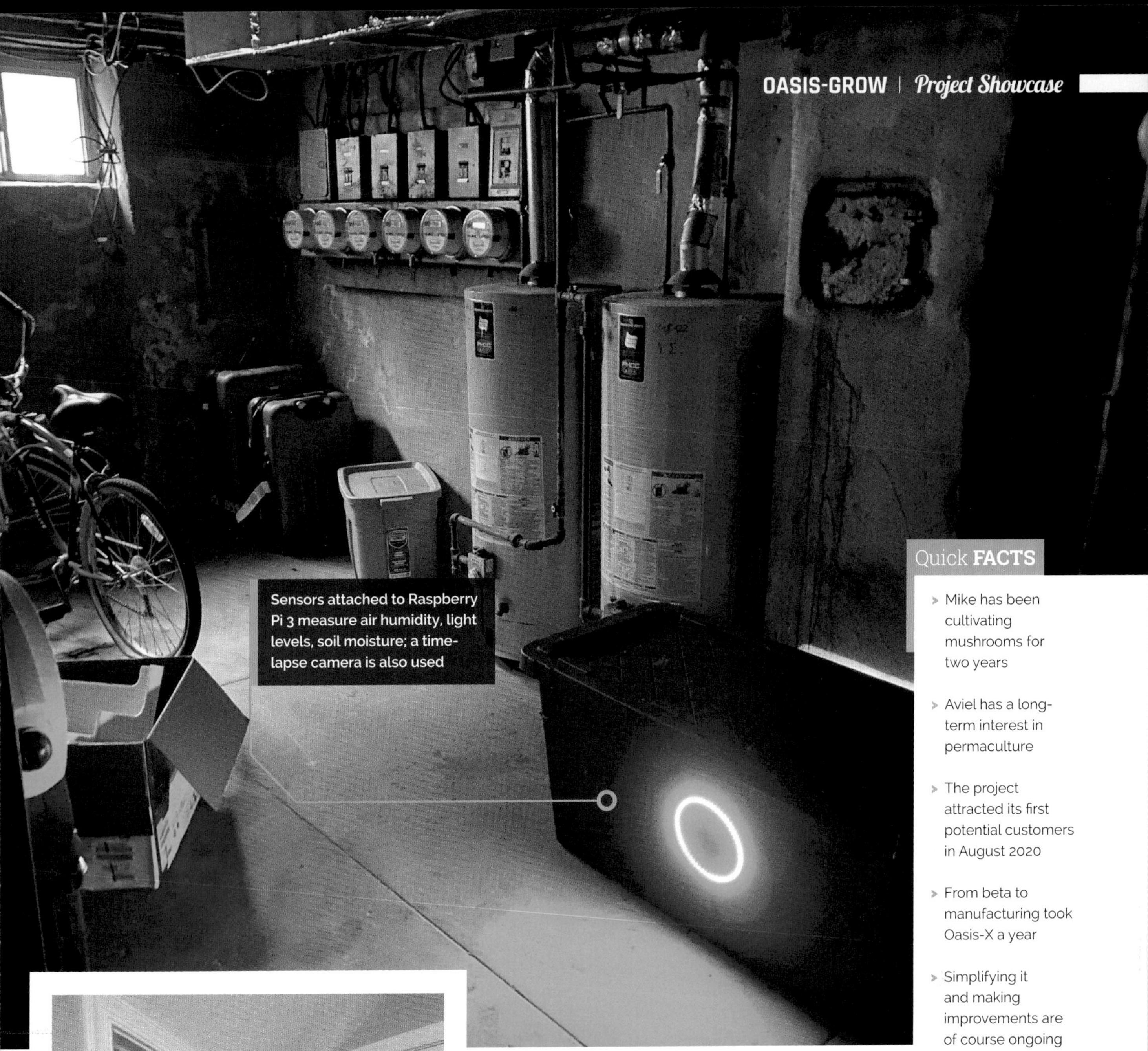

Sensors attached to Raspberry Pi 3 measure air humidity, light levels, soil moisture; a time-lapse camera is also used

Quick FACTS

- Mike has been cultivating mushrooms for two years
- Aviel has a long-term interest in permaculture
- The project attracted its first potential customers in August 2020
- From beta to manufacturing took Oasis-X a year
- Simplifying it and making improvements are of course ongoing

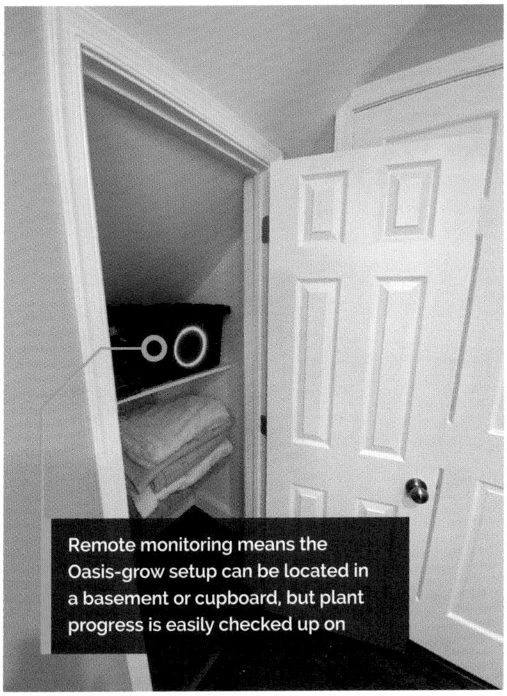

Remote monitoring means the Oasis-grow setup can be located in a basement or cupboard, but plant progress is easily checked up on

Cultures could be cultivated in bottles or jars, or the whole box used to grow mushrooms and other crops

Warning!
Electrical Safety

Please be careful when working with electrical projects around the home. Especially if they involve mains electricity.

magpi.cc/electricalsafety

Time-lapse images can be taken of the contents in their Petri dish habitat

to agriculture, and merit a commercial launch. Mike explains it was important for Oasis-X to take an "open-source first approach" to developing agriculture technology. "Technology vendor lock-in is such a big issue for farmers," he continues.

> Oasis-X is also keen to make the concept available for hobbyists and home horticulturists, providing source code and hardware architecture details via GitHub

"Our systems must collect and store lots of different information while communicating with the network and managing multiple independent processes, [so] a full-fledged operating system is close to required."

Ploughing ahead

Oasis-grow's makers took inspiration from older projects, and chose Raspberry Pi because of its multitasking capabilities and a file system that they found made organising and retrieving complex data a much simpler task. Most of the hardware they use for prototyping was bought from the usual online retailers, with Raspberry Pi running Bash alongside Raspberry Pi OS and Python. Their web interface was created in Python and uses AWS (Amazon Web Services).

"We've spent the last year nailing down the core functionality and are in the process of making it faster, more modular, and easier to use," Mike explains. The system went through more than 20 iterations and was developed in close consultation with farmers and home horticulturists. This process helped Mike and Aviel realise that the vision, sensing, and environmental helper modules were valuable individually as well as

collectively. As a result, the pair are soon set to offer an open-source version of the GUI integrating machine learning, decision support, and smart notification prototypes.

As an agri-business system, Oasis-grow had to prove its worth and has undergone successful long-term, large-scale field tests in farms in the eastern US over the past year. These 'turn-key growing systems' have recently been distributed to the company's early backers, with a Kickstarter campaign fundraising towards its 'smaller helper modules' currently running. However, Oasis-X is also keen to make the concept available for hobbyists and home horticulturists, providing source code and hardware architecture details via GitHub (**magpi.cc/oasisgrowgit**).

▲ Impressive results from their initial home growing chambers

Grow your own

Raspberry Pi 3B, a USB camera, and a power supply are the basis of this project. Full setup instructions can be found at **magpi.cc/oasisgrowgit**.

01 Download the Oasis-grow code from the GitHub page (an installable operating image is in development, take a look at the GitHub instructions). Set up your incubation tent, or position the camera to monitor it.

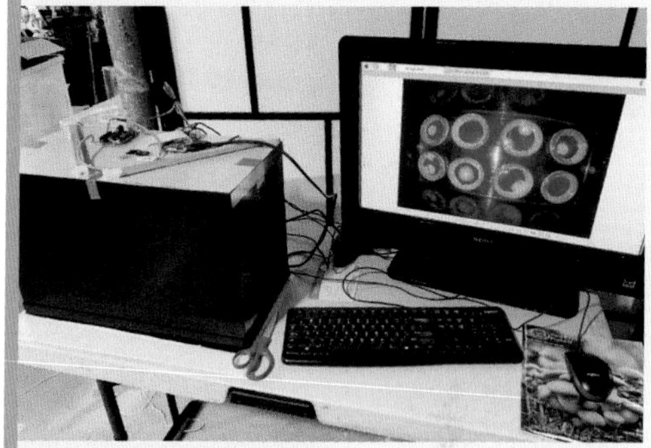

02 To start data collection, turn on the Raspberry Pi camera, start Python, and run **main.py**. If successful, the image loader will run on bootup next time.

03 For more detailed monitoring, add a relay board for AC control, GPIO buttons for a physical interface, environmental sensors, and LED status displays, as per Oasis-grow's full configuration.

Automatic Guitar Tuner

A Pico-powered automatic tool to tune your guitar? **Nicola King** is inspired to start strumming

MAKER

Guyrandy Jean-Gilles

Guyrandy is an engineer and maker from the US east coast. He got his bioengineering degree from the University of Pennsylvania, and currently works in the medical device industry.

magpi.cc/tunergitlab

Let's face it, there's no point in having an ultra-cool guitar if you don't have the means to make sure it stays in tune, especially if you're pursuing a Springsteen-esque form of brilliance on the fretboard.

Guyrandy Jean-Gilles is a musical maker who took Raspberry Pi Pico and created something rather special. "The goal of the project was to make a low-cost, automatic guitar tuner," he tells us. "I'd been meaning to do a project with Pico and there are few development boards as cheap and well-documented."

Pico power

Guyrandy developed his idea over a two-month period, and the result is a very useful device. The guitarist chooses whichever string they want to tune and then places the tuner on the appropriate tuning peg. A button on the automatic tuner is held down by the user which then activates the microphone, and the tuner begins to determine the pitch of the tone coming from the plucked string. It then twists/adjusts the peg as necessary to ensure accurate tuning for sweet sounds.

Guyrandy tells us that Pico is the brains of the entire tuner: "It samples an electret microphone at 4096 samples per second, computes an FFT (fast Fourier transform), finds the strongest magnitude frequency in the audio, then turns a DC motor clockwise or counter-clockwise until the target frequency and the strongest frequency in the audio match."

> I'd been meaning to do a project with Pico and there are few development boards as cheap and well-documented

▲ Guyrandy demonstrates his automatic guitar tuner, here using it to turn the peg for the top E string

Change your tune

As with all electronic makes, this was a learning experience with a few bridges to be crossed, and the trickiest part of the build was determining a guitar string's pitch from audio. "There's a lot of academic research in the area that I wasn't aware of before starting the project," says Guyrandy. "This version found the strongest magnitude in a fast Fourier transform, and assumed that was the fundamental frequency."

In terms of accuracy, he found a few blips that he had to even out. "The tuner's frequency detection is repeatable but inaccurate. It will mistakenly think harmonics are the fundamental frequency for almost all cases. I had to hard-code harmonic frequencies into the firmware to make the tuner work appropriately," he explains.

That said, the tunings are reliably inaccurate so they can be used to correctly tune a guitar. "For example, the low E string is 82.4Hz in standard tuning, but the project repeatable thinks it's 250Hz. Currently alternate tunings aren't possible, but with a firmware change, drop tunings can be achieved."

Guyrandy has had plenty of suggestions from other makers in terms of potential improvements, including adding a vibration sensor so the tuner can be used in a noisy setting, and also "to make a tuner that fits over all the tuning heads of the guitar and tunes all the strings simultaneously.

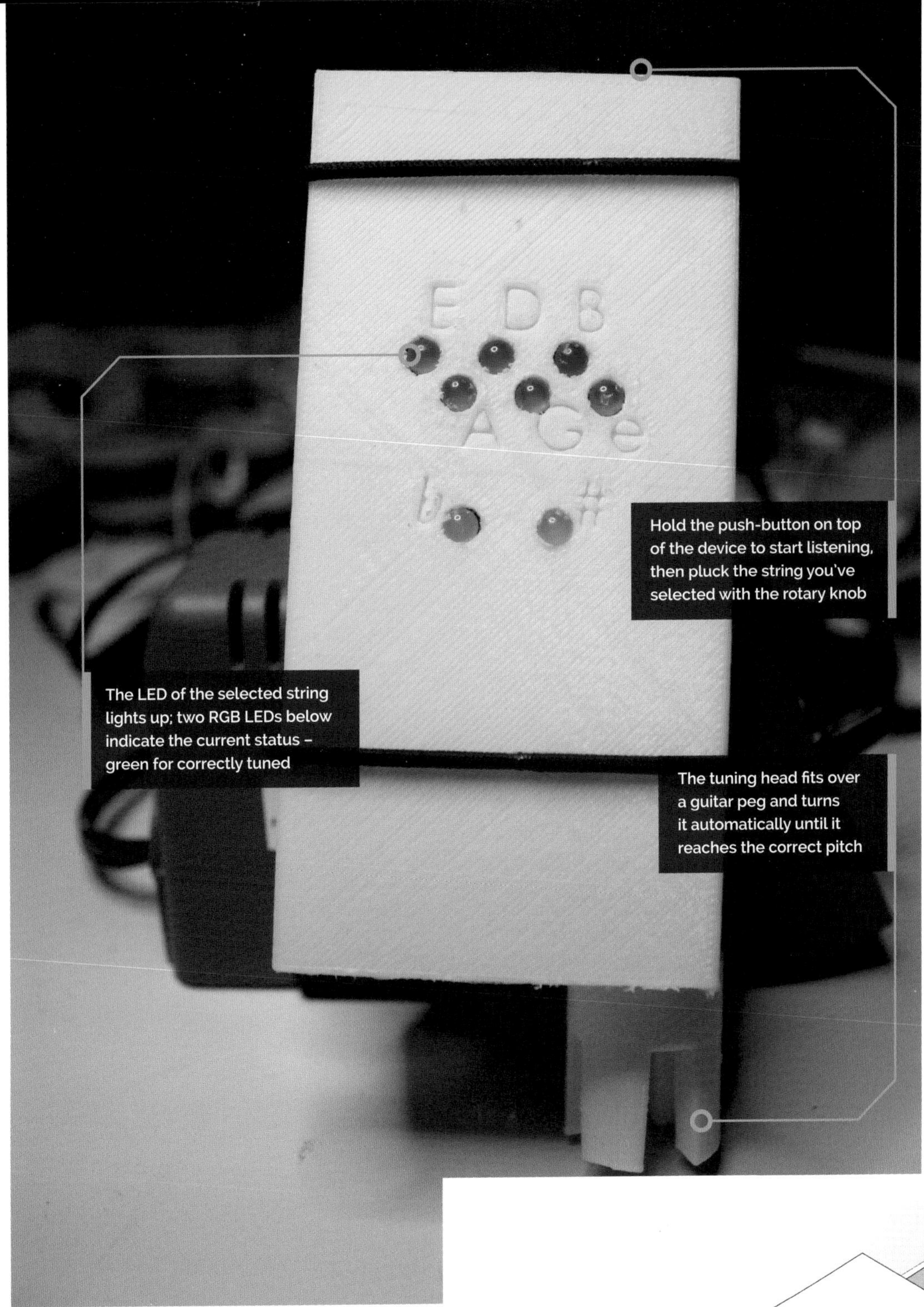
E D B
A G e
b #
Hold the push-button on top of the device to start listening, then pluck the string you've selected with the rotary knob
The LED of the selected string lights up; two RGB LEDs below indicate the current status – green for correctly tuned
The tuning head fits over a guitar peg and turns it automatically until it reaches the correct pitch

Quick FACTS

- Guyrandy was inspired by the Roadie 3 guitar tuner: **magpi.cc/roadie3**
- To make your own, you'll need a Pico, Adafruit PowerBoost 1000...
- ...plus tactile buttons, mini metal gear motor, and some LEDs
- In 2017, Guyrandy snappily decorated his graduation cap and gown...
- ...with the official 7-inch display and a Raspberry Pi 3: **magpi.cc/gradcap**

▼ Guyrandy designed a slick-looking case for the tuner in CAD software and 3D-printed the parts for it

While this is cool, I'm not sure how I'd separate two different fundamental frequencies from one audio signal."

He is currently working on a second version of the tuner and will be trying the YIN (**magpi.cc/yin**) algorithm to detect frequency, as well as incorporating some piezo-electric sensors to pick up vibrations, along with a stronger motor.

If you'd like to have a go at making your own version of the tuner, Guyrandy has generously made his code open-source, and information on exactly what you need in order to build it can be found on his GitLab page (**magpi.cc/tunergitlab**). Why not dust off that old guitar and get making?

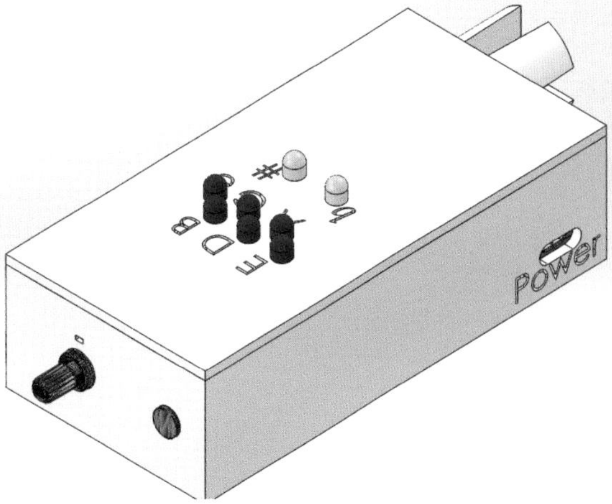
Power

Robot Arm Clock

A deliberately over-engineered timepiece provided a satisfying distraction for a busy software engineer, learns **Rosie Hattersley**

MAKER

Hendrik Ohrens

Full stack software developer Hendrik enjoys playing lacrosse, and making things in his spare time.

magpi.cc/robotarminsta

Like many of us, 33-year-old software engineer Hendrik Ohrens likes to combine the technical skills he uses for work with those he's picked up through his hobbies, to come up with something creative. "I love working on projects that involve electronics, mechanics, and software," he says of his enthusiasm for making things. With a background in scientific research and mechanical engineering at the University of Hannover, plus a few Raspberry Pi builds under his belt, he decided it was time to challenge himself to design his first robot. He decided on a robot arm, since it seemed relatively simple to design but "still has some degree of flexibility". When Hendrik's existing clock stopped working, the idea of using a robot arm to move the dials to the correct position immediately sprang to mind. View the robot arm clock in action on Instagram: **magpi.cc/robotarminsta**.

Monitoring every movement

Hendrik began by using inverse kinematics to work out how to control his robot arm's motion – a scenario that involves using formulae to determine where the end of the arm will swing round to – but soon realised this wasn't the best approach. Instead, he switched to Raspberry Pi 3B+, with the intention of continuing to explore inverse kinematics and the ikpy library alongside AI and computer vision. In the end, he found training the robot to adopt specific poses was the most useful approach for his needs. "The position teaching feature allows me to move the arm to a position (with my hands) and then save that position for further use. This feature turned out to be very useful for the tasks the robotic arm performed after that," Hendrik explains.

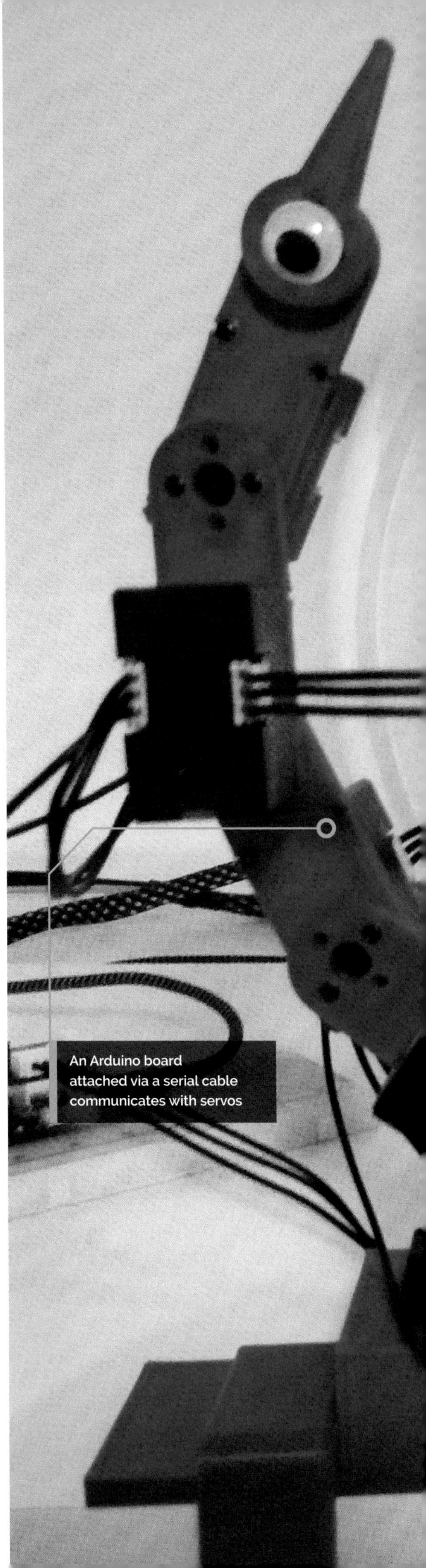

An Arduino board attached via a serial cable communicates with servos

When the clock stopped working, Hendrik decided to use the robot arm to move its hands to show the time

Raspberry Pi 3B+ provides the processing power needed to issue precise position instructions so the robot's arms mimic the hands on the clock

Quick FACTS

- Hendrik allows his projects to take shape organically
- He says the charm of making is to watch things happen when they happen
- But remember to double-check your 3D designs before printing them!
- Hendrik's previous project was also a robot arm
- He trained it to chase a rubber duck

▲ Adding stick-on googly eyes was an instant way of giving the robot arm clock personality

▶ The robot arm moves the clock's minute hand to update the current time

> "Hendrik has not only worked on other Raspberry Pi projects, but has accrued enough of them over the years to possess a dedicated Pi Box to keep them all in"

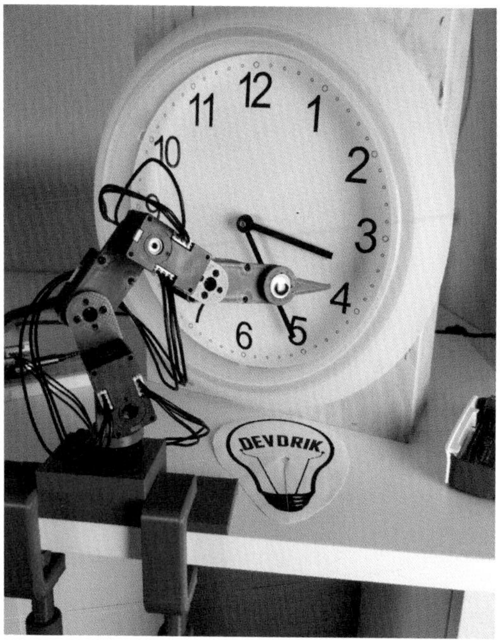

"Raspberry Pi is the perfect development tool for me. It is powerful, versatile, and offers all I need to prototype my projects," he says. Hendrik has not only worked on other Raspberry Pi projects, but has accrued enough of them over the years to possess a dedicated Pi Box to keep them all in. "The fact that I can run modern full stack applications, as well as talk directly to connected hardware, makes it my first choice when I prototype a project."

He decided on an Arduino to talk to the servos using the manufacturer's library, and Raspberry Pi to handle all the logic, as well as controlling the Arduino attached via a serial interface and a USB cable. For this he wrote custom code, refined over time (and still ongoing) to control the arm.

Sketching it out

Hendrik prefers to sketch out plans and adapt them as needed, rather than starting with a rigid design. For his robot arm, he started with some servos and a basic CAD model that would be easy to 3D-print and add small parts to. Once he was

happy with the basic design, he simply duplicated it to make additional movable parts and chained the servos using a cable.

The robot's arm needed to be mounted to prevent it toppling over, so Hendrik 3D-printed clamps he'd found on Thingiverse, and clamped the arm to the table. He then glued the broken clock (whose dials the robot arm would move) to a wooden box, which he also then glued to the table so it wouldn't budge. After some adjustments to the arm's movement, which squeaked because it slightly overlapped in places, Hendrik began the far lengthier process of training the robot which position to adopt when. Having created the somewhat expanded motion-control code, he says the robot arm could be used for any sort of Raspberry Pi project he likes.

▲ Teaching the robot arm which position to adopt involved demonstrating each pose

Arm yourself

01 Download and 3D-print the robot arm parts, including a clamp for the table, if needed. Hendrik used five Dynamixel XL330-M288-T servos, along with an Arduino board and Raspberry Pi 3B+ for his version.

02 Attach and install a Dynamixel Shield and Arduino MKR, then flash dynamixel_api from the **code/arduino** folder to your Arduino. You also need to set up and install Python 3 and ikpy lib on your Raspberry Pi and attach a power supply.

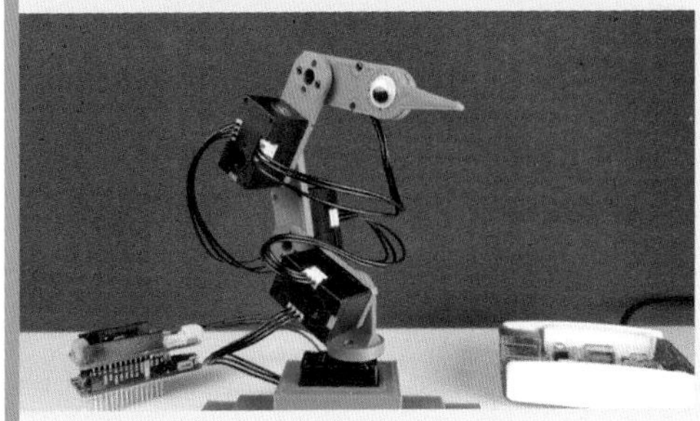

03 Build the robot arm by putting a servo in the base; screw the base_element on to it, followed by the arm_elements. Clamp the base to a table or something solid, otherwise the arm will fall. For complete setup details and code, see: **github.com/devdrik/robo-arm**.

Droiid – A Package Delivery Robot

Here's an autonomous, efficient bot who'll deliver goodies to your doorstep. **Nicola King** waits for her order...

MAKER

Eben Kouao

Eben builds DIY prototypes based on new technologies. From smart mirrors to delivery robots, you can find more of his projects at smartbuilds.io.

magpi.cc/ebenkouao

Inspired by the likes of Amazon and Starship, two companies that have both created six-wheeled delivery robots (magpi.cc/amazonscout and starship.xyz), UK-based Eben Kouao wondered if he could possibly build a similar vehicle. As powerful mini-computers become more accessible, Eben started building his own version with a Raspberry Pi and an Arduino, and the result of his labours is Droiid, a clever robot that can be controlled from anywhere in the world.

Direct Droiid

"Droiid came through the inspiration of recent innovations going on within the delivery market," Eben tells us. "As online consumer demand increases, to keep up with growing demand, the idea of autonomous drones delivering small packages to your doorstep isn't too far off reality."

Furthermore, a Raspberry Pi 4 is central to how this robot operates, as Eben explains: "The robot can be seen as a Raspberry Pi connected to an Arduino attached to motors. Raspberry Pi acts as the orchestrator behind controlling all components of the robot."

In terms of dimensions, Droiid is not huge (330 mm W × 380 mm H × 340 mm L), but it has a compartment big enough to hold small parcels or food deliveries, and it can be controlled remotely via a livestream chat from anywhere in the world.

Droiid is equipped with a mic and speaker for its audio capabilities, including text-to-speech

Eben has also attached a Raspberry Pi Camera Module, a speaker, and a mic module, and so the robot's view of the world can be streamed on YouTube or Twitch. Droiid can be controlled through the livestream chat. For instance, a user can write a command (as a message) in the live chat to Droiid, e.g. 'move? Right'. This command is then sent to a server, and Raspberry Pi receives that message from the server. Raspberry Pi then sends the command to the Arduino, and Droiid receives the message and executes the function.

In addition, as well as giving the robot a command to move, users can use TTS (text-to-speech). So,

A Raspberry Pi Camera Module enables Droiid to see and relay the view to the web interface

Droiid runs on six stroller wheels driven by 12 V high-torque DC geared motors

using the 'say?' command at the start, users could tell Droiid to say 'Hello, today is Thursday.'

The wheel deal

Droiid was built using 3D-printed parts and, due to the size of the project, the printing had to be broken down into smaller parts in order to be accommodated on the 3D printer bed.

Eben decided to equip Droiid with six wheels mainly as a trade-off between the robot's cost and its power. "An earlier model of Droiid used four wheels," he shares. "However, using six wheels seemed to be the sweet spot to get as much power [as possible] from the 12 V geared DC motor to climb inclines with a payload."

To aid stability over rougher terrain, such as grass, an accelerometer was added so the tilt level can be monitored. An ultrasonic sensor gives Droiid spatial awareness and the ability to detect obstacles. Finally, a bottom-mounted WS812B LED strip lights the way for the robot in the dark.

Exciting possibilities

The feedback that Eben has received from all corners has been hugely positive, as there are clearly exciting possibilities regarding how

Quick FACTS

- 3D-printing the various parts at a 40% infill took Eben over a week
- In all, the project took several months to create
- Eben is no stranger to project-building with Raspberry Pi
- Take a look at his Smart CCTV Camera: **magpi.cc/eksmartcctv**
- ...Or his Smart Mirror AI: **magpi.cc/eksmartmirror**

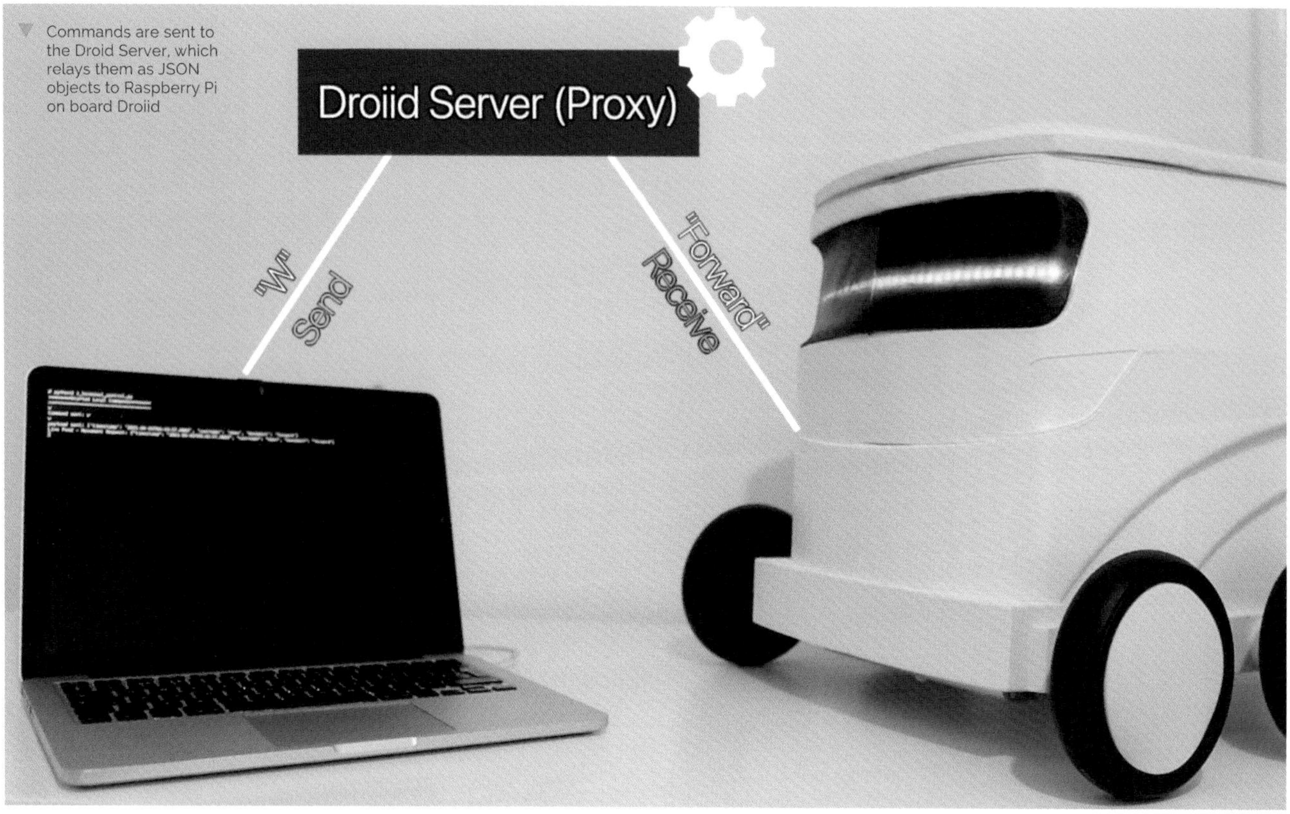
Droiid Server (Proxy)
"W"
Send
"Forward"
Receive

Commands are sent to the Droid Server, which relays them as JSON objects to Raspberry Pi on board Droiid

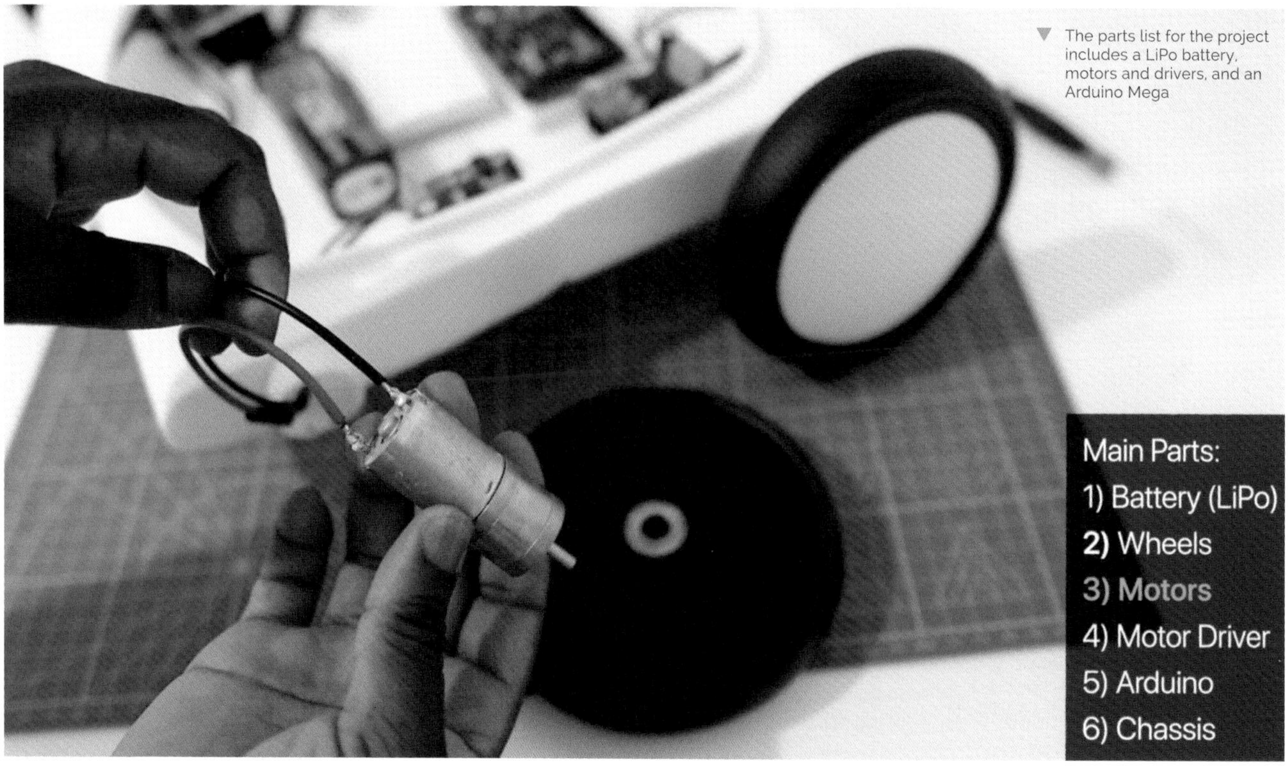
Main Parts:
1) Battery (LiPo)
2) Wheels
3) Motors
4) Motor Driver
5) Arduino
6) Chassis

The parts list for the project includes a LiPo battery, motors and drivers, and an Arduino Mega

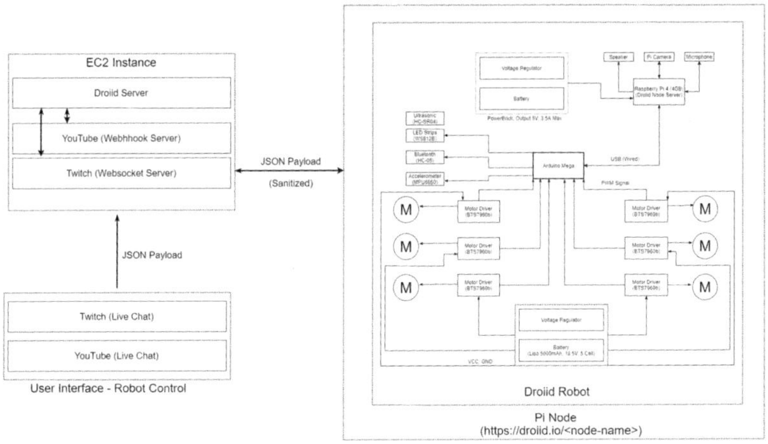

▲ A high-level block diagram showing how all the elements of the project work together

such a robot can be used. "It's definitely a conversation starter," he remarks. "From the maker community, it's also been awesome – some others are curious how it works and what's next. I'm now working to take it out to deliver an actual package!"

> "I'm now working to take it out to deliver an actual package!"

This is just version 1 of Droiid and Eben guarantees that a lot more "features and intelligence" are on the way. Emboldened by the success of the build so far, he is intent on making a number of upgrades. These include improving Droiid's latency using Apache Kafka – an event streaming platform – so that the time taken from a message being sent to Droiid, to the robot actually performing the action requested, is improved.

Eben has also created a couple of additional robot models, including Droiid Mini, which is the four-wheeled version, and is keen to explore the concept of widening the cohort of people who can actually control Droiid's movements. "The idea of others controlling the robot over livestream to complete tasks [or at] sports events is something I'm interested in looking further into," Eben enthuses.

Controlling Droiid

Using the livestream chat in YouTube or Twitch, Droiid can be instructed to move or say something.

01 A user writes a robot command as a message in the livestream chat. This is received by the Droiid Server, parsed as a JSON object, and sent to Raspberry Pi.

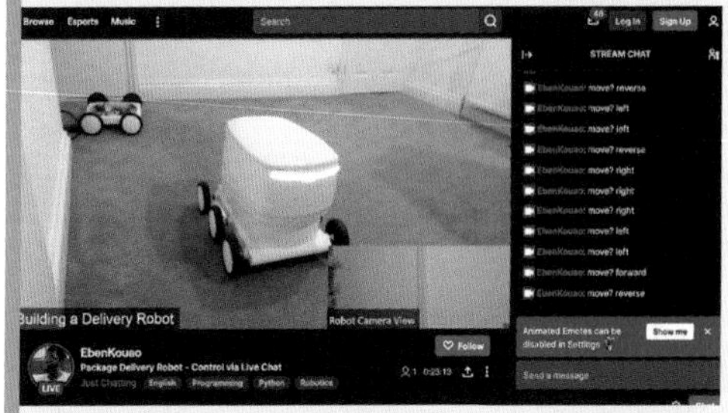

02 If it's a 'say' command, Raspberry Pi sends it to a speaker; if it's a 'move' command, it relays it to the Arduino that controls the robot's motors.

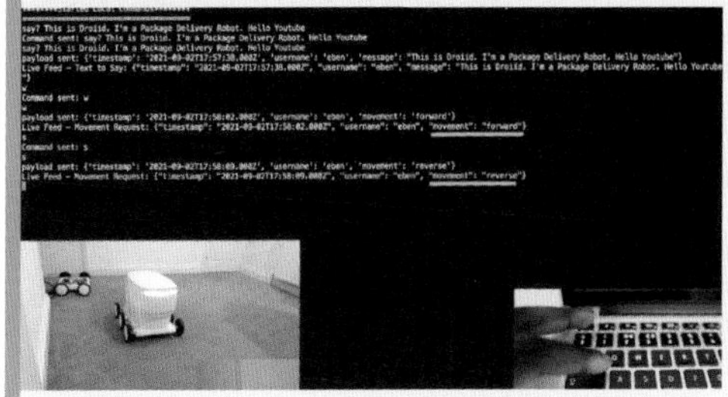

03 Alternatively, Droiid can be controlled manually using the **W**, **A**, **S**, and **D** keys via a remote web interface that includes the view from its on-board camera.

Live CTA Railway Map

This live map of Chicago's 'L' rail system lights up stations where a train is approaching. **Phil King** climbs aboard

MAKER

Jordan von Mulert

Jordan is a violin luthier turned mechanical engineer. His passion is learning and making things. When not with his wife and kids, you'll find him in his workshop or exploring in the woods.

We've seen a few railway-themed Raspberry Pi projects before, including mini departure boards, but Jordan von Mulert has gone one step further by creating a whole railway map – of Chicago's 'L' elevated railway – lit up with LEDs to show the positions of trains at stations in real-time.

Jordan was inspired to make the map after moving away from the Chicago, where he'd lived for ten years with his wife. "Since moving away from the city, we have missed the days of being able to ride the train," he reveals. "Creating this live rail map gives a real sense of connection to the ebb and flow of the city. I like waking up early in the AM and seeing maybe only a dozen trains running in the city. But as the city comes alive, more and more trains light up my map."

Light up the board

Mounted on a wooden board, the map features no fewer than 191 LEDs – one for each station – wired individually to pins on eight 24-channel PWM LED driver boards controlled by a Raspberry Pi Zero W.

Raspberry Pi is at the heart of the operation, as Jordan explains. "On startup it runs my Python script which pulls data from the CTA API about every seven seconds. I found that more often than that doesn't capture any useful changes, but less often than that and you tend to get lots of changes at once."

The CTA's API gives him a complete status of all the trains operating in the system. "When a train approaches a station, the API includes an 'approaching station' flag for that train," says Jordan. "My code looks for these flags and activates the LED mapped to the corresponding station."

The script will also keep track of each train along its run so that it knows when to turn off the LED for the last station and illuminate the LED for the new approaching station, and also to know when a train has left service at the end of the run, otherwise the last LED in the run would be perpetually lit.

Each station on the map has an LED that lights when a train is approaching it; the live data is fetched using the CTA's API

Web of wiring

While Raspberry Pi is the brains of the operation, Jordan tells us the eight LED drivers are the brawn of it, as they control each individual LED. "I debated for a long time as to the best way to obtain independent control of the 191 LEDs needed for the map," he recounts. "Addressable LEDs were an option, but finding 3 mm discrete through-hole

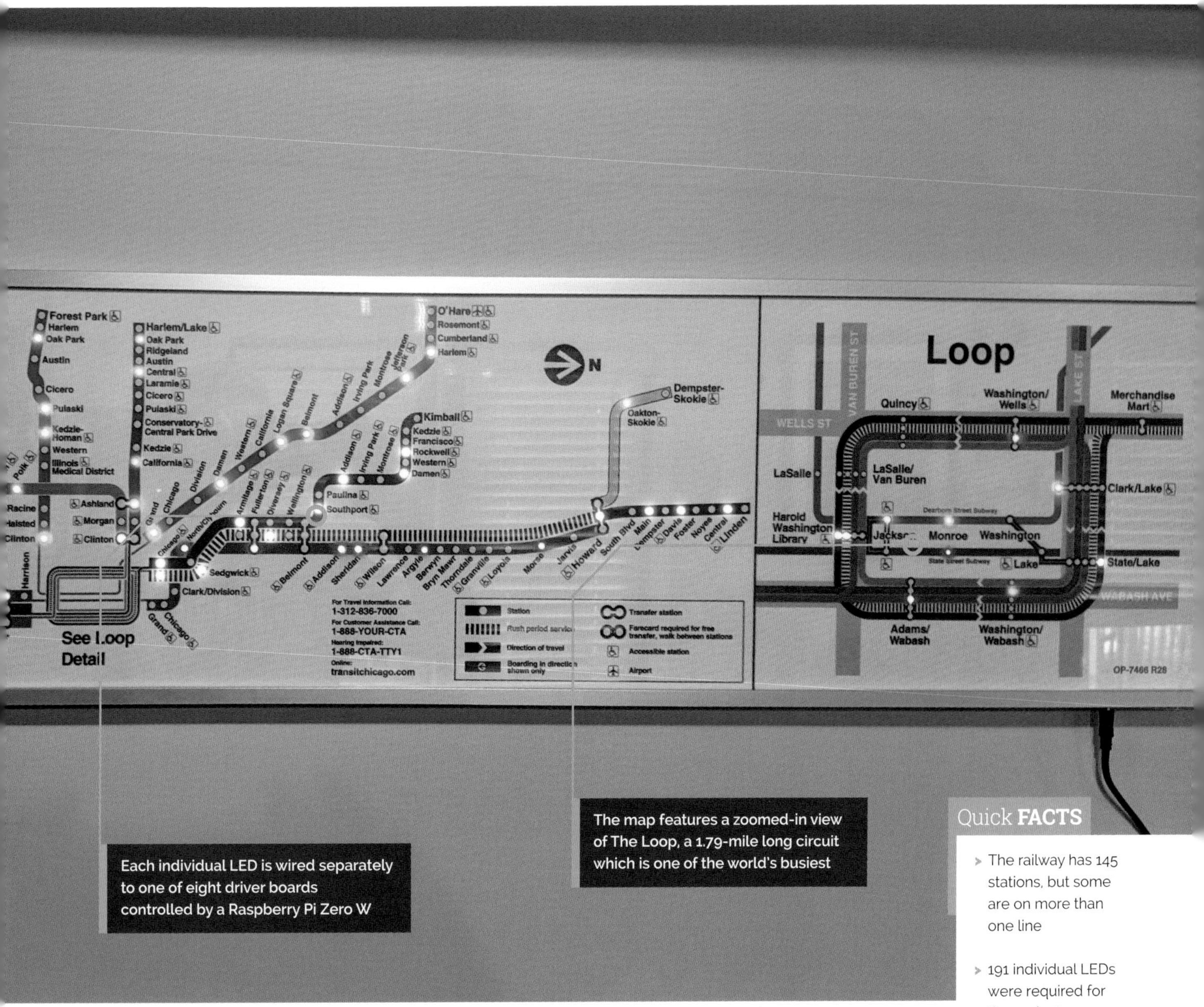

Each individual LED is wired separately to one of eight driver boards controlled by a Raspberry Pi Zero W

The map features a zoomed-in view of The Loop, a 1.79-mile long circuit which is one of the world's busiest

Quick **FACTS**

- The railway has 145 stations, but some are on more than one line
- 191 individual LEDs were required for the project
- Using a custom PCB would have been very expensive
- The physical build only took 12 hours...
- But Jordan spent months brainstorming ways to ease assembly

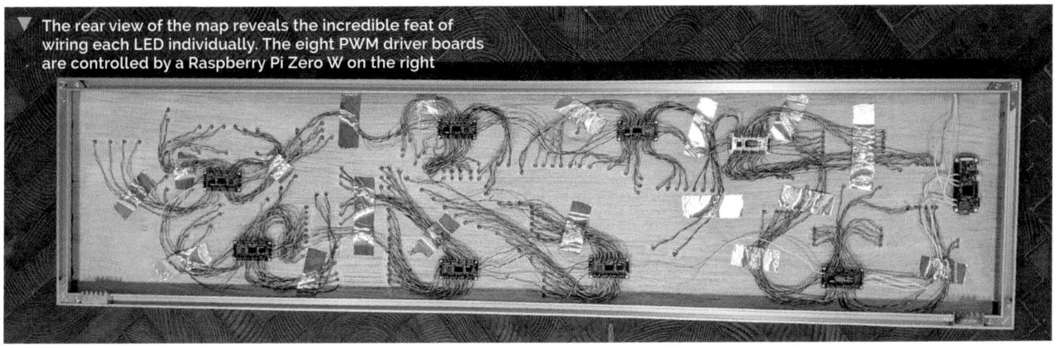

▼ The rear view of the map reveals the incredible feat of wiring each LED individually. The eight PWM driver boards are controlled by a Raspberry Pi Zero W on the right

▲ A close-up of Raspberry Pi Zero W mounted on the rear of the map, showing the wire-wrapped connections to its GPIO pins

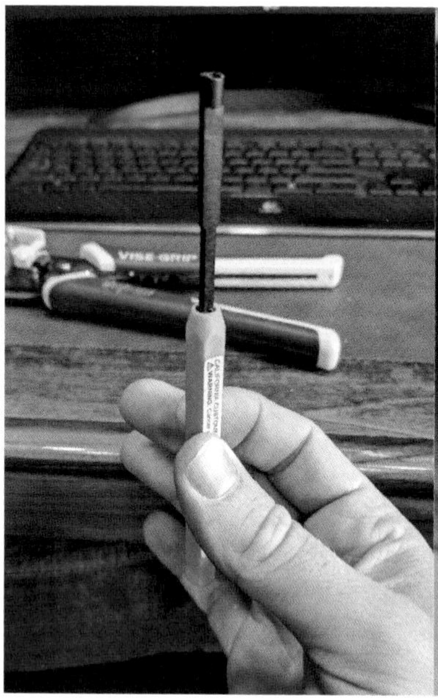

▲ Jordan used a special tool (left) to connect all the LEDs by wrapping wire around their legs (right) and the pins on driver boards

▼ And so it begins... wire-wrapping connections to a couple of LEDs on one of the eight PWM driver boards – only 189 more to go!

addressable LEDs that could be mounted without a PCB proved to be a challenge. In the end I stumbled across these drivers and determined that they would be workable."

> " As the city comes alive, more and more trains light up my map "

With each LED needing to be connected to one of the driver boards, the project required a phenomenal amount of wiring. Rather than soldering them all, Jordan opted to use wire wrapping, which he says is a lot quicker. "It's so fast and easy with a tool... and they are super secure! Apparently they used wire wrapping on the Saturn V [rocket] – not to suggest that my wraps were as good as NASA, but in principle they can take a shocking amount of vibration. The other benefit to them is serviceability. I can undo a wire wrap to swap an LED in less than the time it would take my iron to heat up."

In the loop

The zoomed-in view of 'The Loop' section of the CTA railway caused a little extra complication, and also precluded using off-the-shelf addressable RGB LED string lights for the project. "Those have 5 mm LEDs which would have been too big for the tight spacing required at each train stop," notes Jordan. "I had considered use different LEDs for that section only, but I wanted the lighting to be consistent."

Jordan spent hours debugging code and mapping each LED channel to the correct stop. "Electrically, things went pretty smoothly," he says. "I ended up adding some optional filter capacitors to the boards to keep electrical switching noise down, and I'm still hunting some gremlins in the SPI data lines. But you know how it is: the reward is the journey, not the end product."

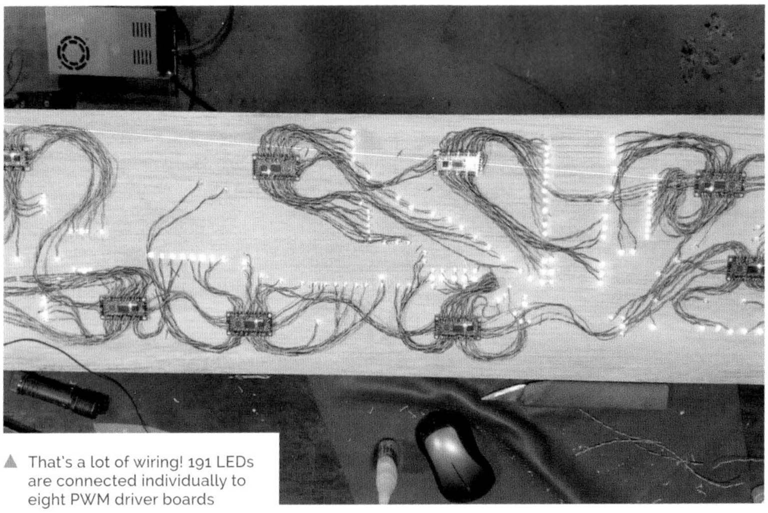

▲ That's a lot of wiring! 191 LEDs are connected individually to eight PWM driver boards

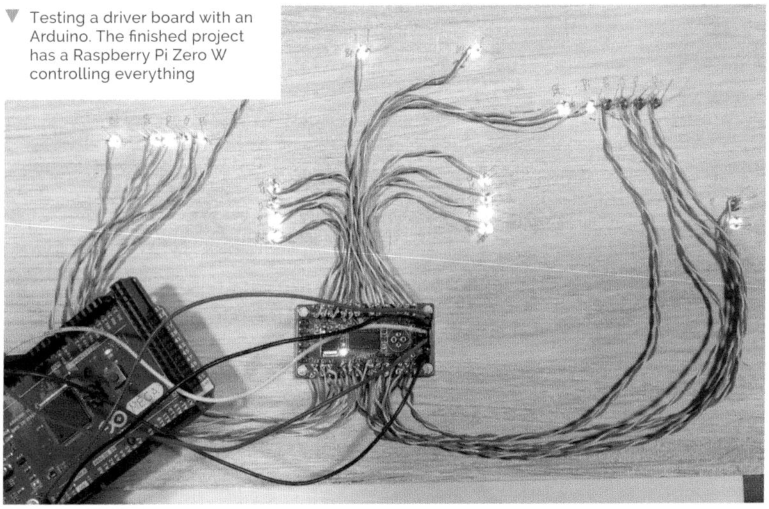

▼ Testing a driver board with an Arduino. The finished project has a Raspberry Pi Zero W controlling everything

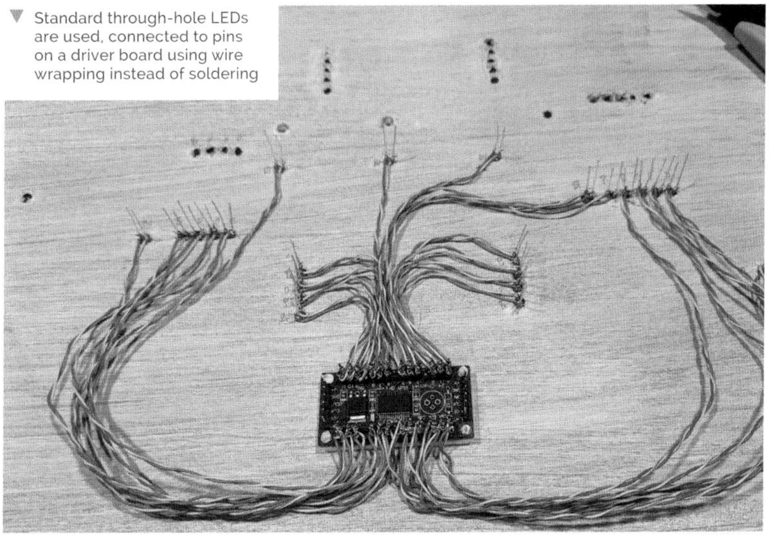

▼ Standard through-hole LEDs are used, connected to pins on a driver board using wire wrapping instead of soldering

Making a map

Creating a railway map with holes for the LEDs to light it up.

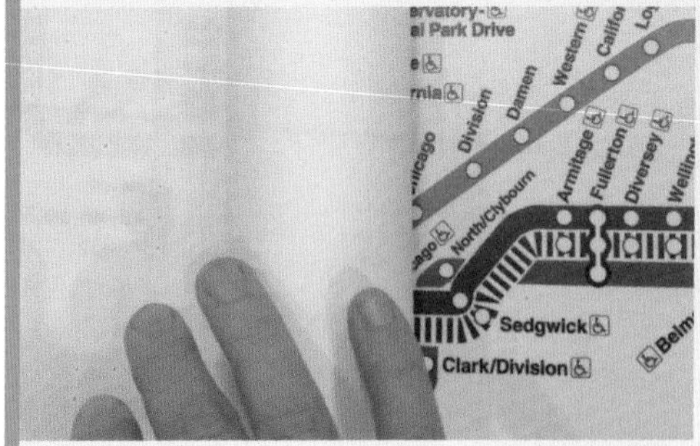

01 Jordan bought a translucent map of the CTA rail system and traced the position of each station on a paper template placed over it.

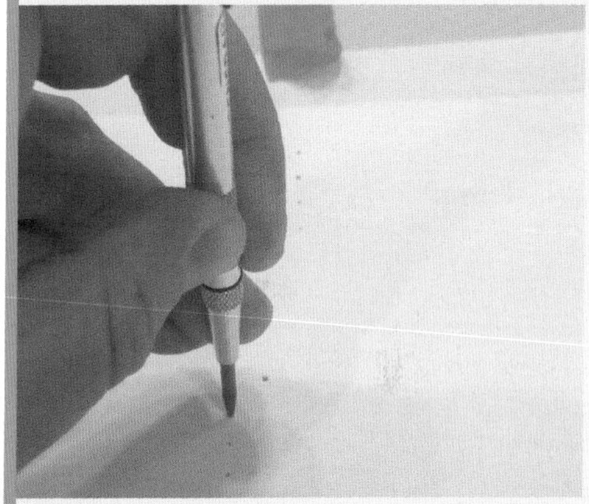

02 He then transferred the paper template to the wooden mounting board and punched a hole at each station for a 3mm LED to fit into.

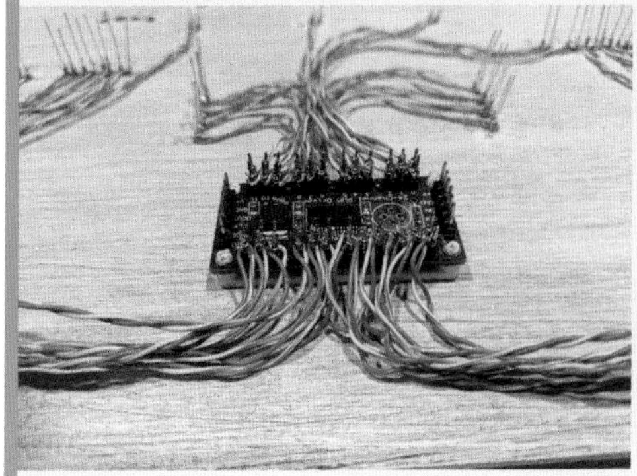

03 Instead of soldering, wire wrapping was used to connect each individual LED to pins on one of the eight PWM LED driver boards.

Automatic Dog Ball Launcher

Are your games of fetch with your pooch leaving you panting? **David Crookes** catches up with a Raspberry Pi Pico solution

MAKER

Brankly

Brankly is an inventor and entrepreneur. He grew up in Germany and worked as an IT consultant for years before following his passion for inventing full-time.

magpi.cc/dogball

Dogs love to play fetch and it's a wonderful way to exercise your pooch. Trouble is, constant ball throwing isn't half exhausting for human arms, which is perhaps why there's a growing number of automatic ball launchers on the market.

Rather than buy one of those, however, seasoned maker Brankly has created one of his own. "My dog loves to play with blue toy balls and I wanted him to have a little more fun," he tells us. "I also want him to be able to play fetch when we are busy by encouraging him to load the balls himself."

While training his dog to do so is still ongoing, work on the ball launcher is complete. It makes use of a Raspberry Pi Pico microcontroller board which is something of a departure for Brankly who has previously used Arduinos for his projects. "I just wanted to try something new," he says. "I love how simple it is to program."

▲ Brankly says the limitations of his 3D printer meant the components had to be cramped into the case more than he'd have liked

Print perfect

Before getting down to coding, however, he spent time experimenting. "I started with the motor mount to test if that would launch the balls," he explains. "After that, I designed the other parts and, in the end, the case. I played around with different shapes and found that a sphere design looked the best. It was also small enough for my 3D printer to print."

The case was designed in Fusion 360. "I like 3D printing because you can design everything first, send it to the printer, and have the part after some hours," Brankly says. "Most of the time, you have to change small parts, but it's a pretty straightforward process." Indeed, he refined the project as he went along, such as the rollers that accelerate the ball.

"The first version had a profile printed on them, but the friction was not enough to grab the ball when it was wet from my dog's saliva," Brankly says. "I also had to change the mount for the sensor a couple of times because the first two sensors I tried didn't work."

And fetch

So how does the launcher work? When a ball is placed into its funnel, it is prevented from falling into the launch channel by a piece of plastic that's controlled using an SG90 servo. The ball is then detected by a sensor, prompting Raspberry Pi Pico to get ready for launch.

"I used a motor controller to be able to randomise the speed of the motors with a pulse-width modulation (PWM) signal," explains Brankly. "For each cycle, Raspberry Pi Pico generates a random number between 40,000 and 65,000, and this gets sent to the controller as a PWM value.

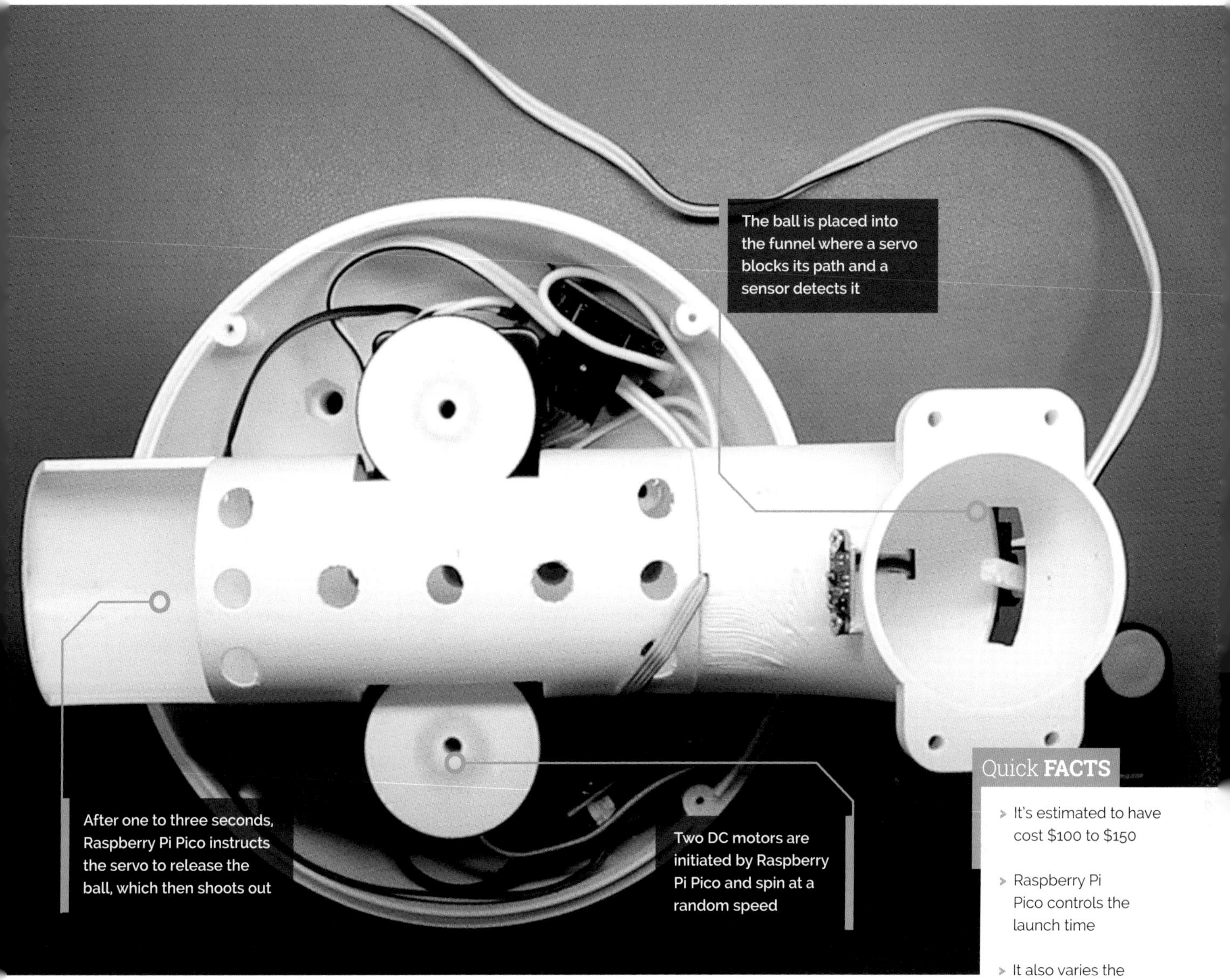
The ball is placed into the funnel where a servo blocks its path and a sensor detects it
After one to three seconds, Raspberry Pi Pico instructs the servo to release the ball, which then shoots out
Two DC motors are initiated by Raspberry Pi Pico and spin at a random speed

Quick **FACTS**

- It's estimated to have cost $100 to $150
- Raspberry Pi Pico controls the launch time
- It also varies the motor speed
- Make sure to use soft balls
- Print files are on Thingiverse: **magpi.cc/dogballthingiverse**

"I used a motor controller to be able to randomise the speed"

"This will vary the distance of the ball each time so it's more fun for my dog. After the motors are started, the ball releases and gets shot out. The motors turn off, the servo blocks the entrance again and the machine is ready for the next ball."

The approach adds an element of uncertainty and means a dog won't know exactly when the ball is going to launch, or how far it's going to travel. "I'm really happy with the end result," Brankly says. "I think the design looks pretty cute and my dog really loves it. He gets all excited when he hears the machine starting up."

▼ The motor controller can be seen to the left, with the motors themselves in the centre of the image

Old School Minitel Laptop

From Minitel to a near-mini laptop, Gautier Serodon has repurposed a terminal once used for France's innovative online service in the 1980s, as **David Crookes** explains

MAKER

Gautier Serodon

Gautier is an engineering student, as well as an electronic and DIY enthusiast. He enjoys retro tech, vintage cars, watchmaking, 3D printing, and board games.

magpi.cc/minitel

Prior to the development of the World Wide Web, France had a hugely popular telecommunications service called Minitel. It allowed the country's citizens to book train tickets, check their electronic mail, search the telephone directory, and access online banking among other things, attracting an estimated 25 million users and offering around 26,000 services at its peak.

Launched in 1982 and remaining in use for exactly 30 years, it was far ahead of its time. Anyone who wanted a terminal to connect to Minitel could get one for free from what became France Télécom, and this led to 9 million sets being installed in homes by 1999. But since Minitel closed, many have ended up being sold. "It's easy to find a terminal on sale for below €10," says French maker Gautier Serodon.

Having snapped one up himself at a garage sale, the 25-year-old decided to bring it back to life. "I love retro tech and I wanted to revive my Minitel with today's technology," he says. He decided a Raspberry Pi 3B computer would enable him to do just that. "It's affordable, compact, and internet-compatible," he explains. And the plan? To convert it into a battery-powered laptop so he could take notes during his engineering school classes.

▲ It may not the sleekest of 'laptops', but his charmingly bulky piece of retro kit is still portable

Key to authenticity

Gautier bagged himself a Minitel 1B terminal, made in 1982 by Telic Alcatel. He stripped it of most of its parts, including the CRT display, but decided he wanted to at least retain use of the terminal's original – and satisfyingly clicky – AZERTY keyboard.

He also kept the power button and power indicator. "I wanted to keep as much of this charming old technology as possible," he tells us. "The power button and the LED played an important role in maintaining the aesthetics of the Minitel."

The screen was replaced by a 10.4-inch LCD panel which came complete with a driver board and this was fixed within the Minitel casing using four 3D-printed parts, one for each corner. The biggest challenge, however, involved retrofitting the old Minitel keyboard so that it would simulate a USB HID keyboard that Raspberry Pi could work with.

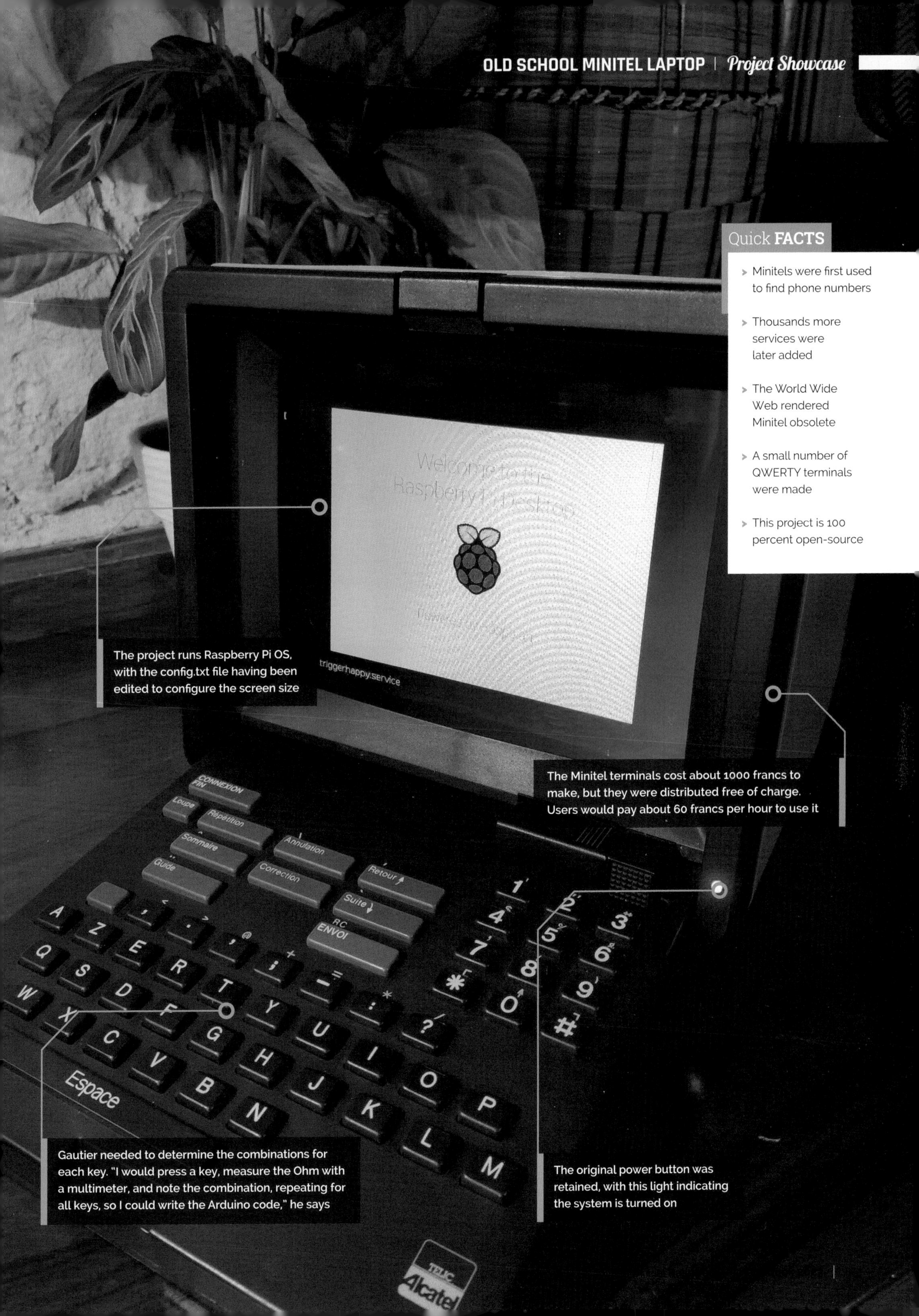

Quick **FACTS**

- Minitels were first used to find phone numbers
- Thousands more services were later added
- The World Wide Web rendered Minitel obsolete
- A small number of QWERTY terminals were made
- This project is 100 percent open-source

The project runs Raspberry Pi OS, with the config.txt file having been edited to configure the screen size

The Minitel terminals cost about 1000 francs to make, but they were distributed free of charge. Users would pay about 60 francs per hour to use it

Gautier needed to determine the combinations for each key. "I would press a key, measure the Ohm with a multimeter, and note the combination, repeating for all keys, so I could write the Arduino code," he says

The original power button was retained, with this light indicating the system is turned on

Warning!
CRT

Be careful with projects involving old television and CRT equipment. Opening up a CRT can be dangerous, risking electric shock even if the TV is not plugged in.

magpi.cc/crt

▶ Here you can see the internal wiring, connecting the screen (laid flat) to Raspberry Pi fixed to the right of the case

▼ An HDMI cable connects Raspberry Pi to the display. You could easily upgrade to Raspberry Pi 4

▲ Given it dates back to 1982, Gautier's Minitel terminal – bought from a garage sale – is in surprisingly good condition

```
       //  5   4   3   2  15  14  13   9
/* 6  */{'c','D','Q','S','r','l',' ',' '},
/* 7  */{'L','F','Y','A',',','r','s',' '},
/* 17 */{'K','G','Z','T','.','a','c',' '},
/* 16 */{'J','H','E','R','"',' ','g',' '},
/* 12 */{'V','C',';','U','e','r','s',' '},
/* 11 */{'B','X','-','I','*','4','7','1'},
/* 10 */{'P','W',':','N','O','5','8','2'},
/* 8  */{'M',' ','?','O','#','9','6','3'}};
```

▲ This is the keyboard matrix Gautier used, with eight columns and eight rows

> It's possible to plug an Ethernet cable directly into Raspberry Pi too

"The Minitel keyboard is a matrix with a 17-wire output cable," Gautier explains. "When you retrofit a matrix keyboard, the goal is to find the wires of the columns and the wires of the rows – in this case, there are eight rows and eight columns, making a keyboard with a maximum of 64 keys."

Looking rosy

Gautier used an Arduino Pro Micro which has an ATmega32U4 microcontroller on board. "This is the easiest way to emulate a keyboard," he says. "Other boards such as the Arduino Nano don't have the same microcontroller, so it's way more difficult to simulate a keyboard without modifying the bootloader and other things."

From this point, the main work involved fixing a Raspberry Pi in place and connecting it to the keyboard, a DC step-down converter, and the power switch. "I placed Raspberry Pi right behind the Minitel's hatch [located behind the screen] so that it is easy to plug in a dongle for a wireless mouse," Gautier says. "It's possible to plug an Ethernet cable directly into Raspberry Pi too."

All in all, it makes for a very neat setup, one that is powered by a 12 V Li-ion rechargeable battery for portability. Gautier doesn't attempt to emulate the UI of an old Minitel system – "I'm simply running Raspberry Pi OS," he says – but it still gives a flavour of what it was like to use the system. "It's just a shame I never had the chance to use a Minitel for its original applications," he laments.

Mini-guide to revitalising a Minitel

01 The bulk of the electronics are behind the Minitel's screen and need to be removed. A new screen is fixed using printed parts. STL files of these are provided by Gautier.

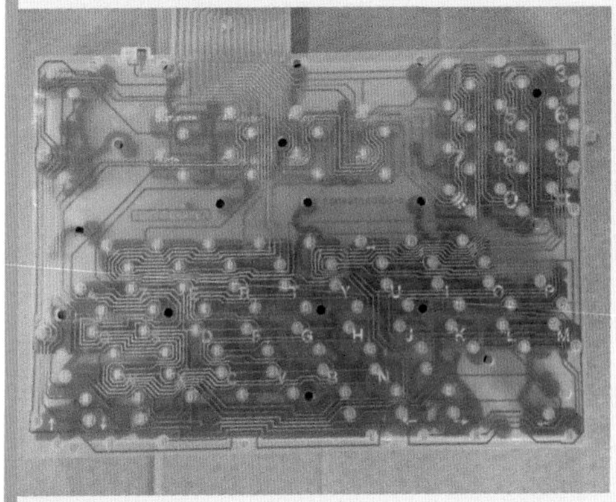

02 A Minitel contains a simple matrix keyboard. The matrix is constantly scanned, allowing it to detect when a key is pressed and a circuit is closed. It must be mapped.

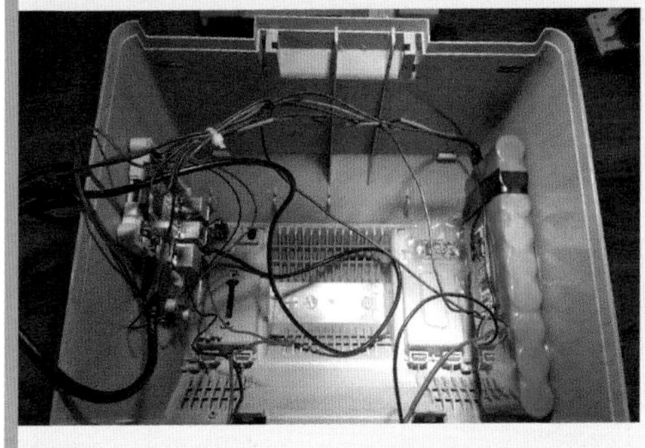

03 Now the case can be kitted out with Raspberry Pi (in a 3D-printed case), a battery, a light indicator, and power supplies. Hot glue is used to affix the components.

Big Boxes

Social enterprise Jangala provides internet access to communities affected by disasters. **Rosie Hattersley** hears their big idea

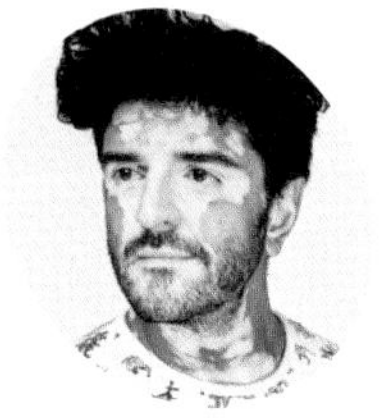

MAKER

Rich Thanki

The possibilities of wireless connectivity led community-minded Rich, a self-confessed electronics tinkerer, to co-found a non-profit helping others get online.

janga.la

For communities affected by man-made or natural disasters, web access provides a critical lifeline.

Not-for-profit organisation Jangala provides internet access for displaced people. The organisation was born out of its founders' involvement in humanitarian relief efforts. These included the Amatrice earthquake in central Italy in 2016 and supporting refugees living in Camp de la Lande, Calais.

Jangala's ruggedised Big Boxes provide web access in challenging environments and, with a recent Raspberry Pi upgrade, are now capable of offering connectivity to multiple users via almost any wireless means. Their mission is to connect every school, clinic, and community resilience project worldwide. In 2019, the project won a Tech4Good Africa award, having supplied internet access to 50 students at a refugee camp in Kenya, enabling them to complete their diplomas.

▼ Each box provides secure wireless access, weighs just 1.9 kg, and fits in a small backpack

The Big Easy

Jangala co-founder and head of technology Rich Thanki has been fiddling with gadgetry and taking electronics apart "as long as I can remember". A huge fan of wireless technology, Rich worked on projects with Microsoft bringing connectivity to people in sub-Saharan Africa before founding Jangala with Samson Rinaldi, who managed the CAD and hardware design aspects.

Each Big Box is a briefcase-sized device, designed to be simple to set up and take advantage of an available wireless connection to offer WiFi access to hundreds of users at a time. "Scalable high-quality WiFi without the need for costly technical expertise" characterises Jangala's single-box approach to providing internet access when and where it may be needed.

This is the first of Jangala's projects to use Raspberry Pi products. "I think the first of many," says Rich.

Jangala's updated purple box is now faster, more capable, and more rugged, the team explains, with the wireless antennas having been moved inside the box to protect them from the elements, plus twin modems that allow Big Box to connect to networks ranging from 2G networks to 5G. It was largely this aspect that prompted the switch to Raspberry Pi since it can handle the gigabit speeds that 5G networks offer.

The ruggedised purple boxes are briefcase sized and stuffed with enough electronics to connect hundreds of users wirelessly

Big Boxes support gigabit wireless and can be booted and updated remotely

Upgrading from a previous system to Raspberry Pi resulted in far more power, the need for a cooling system, and the ability to connect to anything from 2G to 5G

Quick **FACTS**

- Jangala has set up 20 Big Box installations so far
- They provide web access to more than 25,000 users
- Jangala's Get Box provided UK users with web access during lockdown
- Each Big Box has a 4G global modem
- It can also connect via satellite

▲ Deployments include refugee camps such as this one for exiled Rohingyans
Credit: DFID, Rakhine camp, Myanmar

◀ Delivery of Big Boxes at Kakuma refugee camp, Kenya, where they will help support women and disadvantaged youths gain education and employment

Other components were sourced from "absolutely all over the place", but the use of OpenWRT and the support of the Linux community were absolutely crucial. Rich also credits amazing manufacturers in China who designed the incredibly useful things that can't be found anywhere else, "everything from the perfect boost converter to little waterproof slots for SIM cards".

> Without the stable base provided by Raspberry Pi, things would've gotten hairy quickly

Tech for good

"The role of Raspberry Pi in Big Box is essentially to run the show," says Rich. "It's really the brains and brawn of the system. We are able to control and see Big Boxes in the field by talking to Raspberry Pi. It routes every bit of traffic, which is no easy feat. We've a unique software setup to ensure security and it's remotely updatable, so we have to use U-Boot as a second-stage bootloader. Without the stable base provided by Raspberry Pi, things would've gotten hairy quickly."

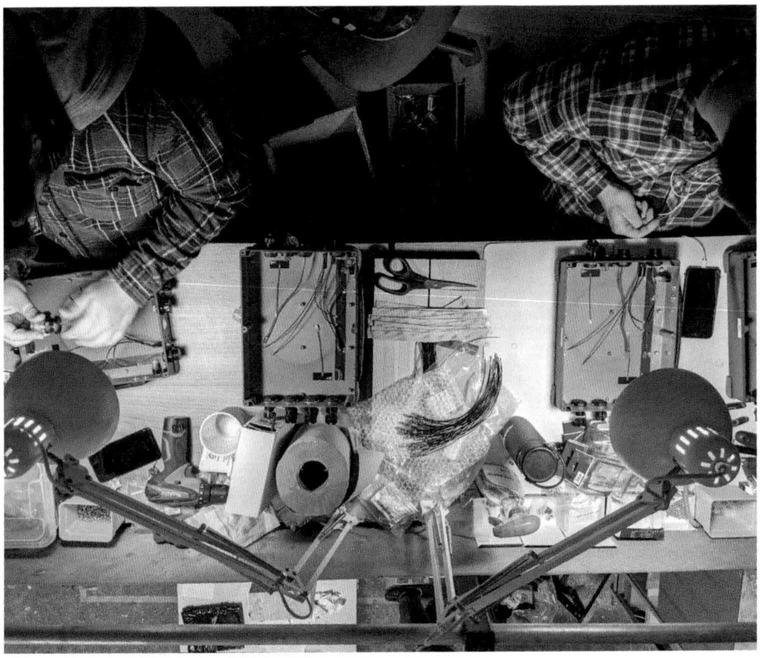

Rich says working with Raspberry Pi 4 has been great because of the community, really accessible documentation, and standards that are open-source. "It's made a huge difference for us. We wouldn't have been able to do this on any other device," he notes.

Although Big Box is Jangala's first Raspberry Pi experience, it won't be their last. "Our next two projects will tackle other aspects of infrastructure which are really important in humanitarian settings, mainly power and edge computing. In both these cases, Raspberry Pi provides things that could easily be at the heart of these projects."

For anyone else planning on designing products for social good, Rich says, "Focus on what you have to do to get the design into the hands of people and making a difference as soon as possible. This might run against the grain of people who want to be perfect and neat. Don't let the perfect be the enemy of the good."

▲ Big Box and Lenovo devices support online education courses in Kalobeyei settlement, Kenya

▲ The boxes provide internet access to aid workers, refugees, and remote communities such as this one in Kenya

Big Box basics

01 Each Big Box provides internet access for more than 100 users, with additional connections added via extra radio controllers. Raspberry Pi provides the "brains" of each box.

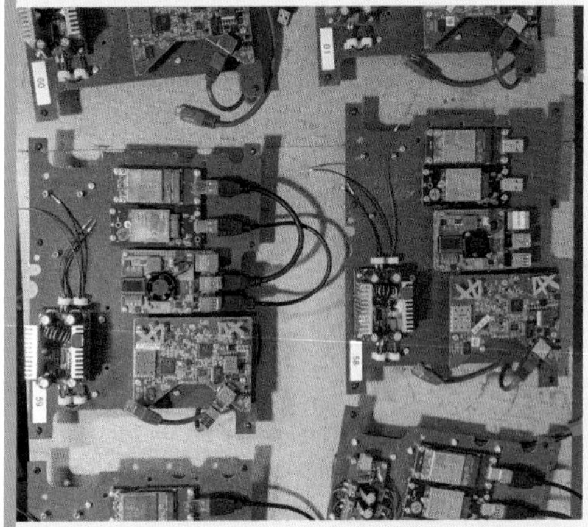

02 The setup includes a PoE (Power over Ethernet) switch, a 4G global modem, and wireless access point, plus bespoke, highly efficient power management.

03 Due to the demanding locations in which Big Boxes are deployed, the ruggedised boxes need to be fully waterproof, including their connectors and 'glands'.

DeMoor Orrery

A 240-year-old orrery provided the inspiration for Chris de Moor's stunning planetarium project. **David Crookes** takes a look

MAKER

Chris de Moor

Former programmer and enthusiastic maker, Chris founded and ran a digital agency called Est Digital until earlier this year. His website links to the required GitHub files.

demoor-orrery.com

Many people have a keen interest in astronomy, but before Chris de Moor created a project that's truly out of this world, he didn't count himself as being among them. "I didn't know anything about astronomy and that sort of stuff," he freely admits. Yet after visiting the Eise Eisinga Planetarium in Franeker in the Netherlands, he was inspired.

"Eise Eisinga completed an orrery in his house in 1781," Chris says, of the oldest working orrery in the world, created by the Frisian amateur astronomer. "I saw it on the ceiling and thought I wanted one just like that, thinking it would fit perfectly in my living room. At first I believed it would be a weekend project – a simple art piece – until I decided it should work as well. And then the rest came." Indeed, what followed would take him a year!

Chris set about building a replica, one that is similarly coloured and with planets fitted to copper tracks. But while Eisinga's orrery was driven by a pendulum clock driving a host of mechanics in the space above the ceiling, Chris went one better: he used six Raspberry Pi Zero computers – one for each of the six planets he decided to concentrate on.

The only departure from scale are Jupiter and Saturn. They should be the size of footballs but Chris made them smaller for artistic reasons

Planetary planning

"My first question was about scale and how big my orrery should be," Chris recalls. Ultimately, he worked out that he only had room for Mercury, Venus, Earth (plus the Moon), Mars, Jupiter, and Saturn. The next step was then figuring how spaced apart they should be, the size of each representative planet and how they would move into position in real-time.

"I'd heard about Raspberry Pi computers so I decided to experiment with them," Chris says. "I also know how to program and, while Python is not my favourite language, it was easy to learn. What I didn't know was how stepper motors worked. They were new to me. But once I bought one Raspberry Pi Zero and a stepper motor and played around, I realised what was possible."

Even so, there were lots of technical challenges along the way. "I got some plywood and started sawing, but a lot of things went wrong," he laughs. "Trying to saw perfect circles by hand is impossible, so I had to go to a store that had a CNC cutting machine, which meant I needed a drawing and DXF vector file. I ended up in areas where I didn't know anything."

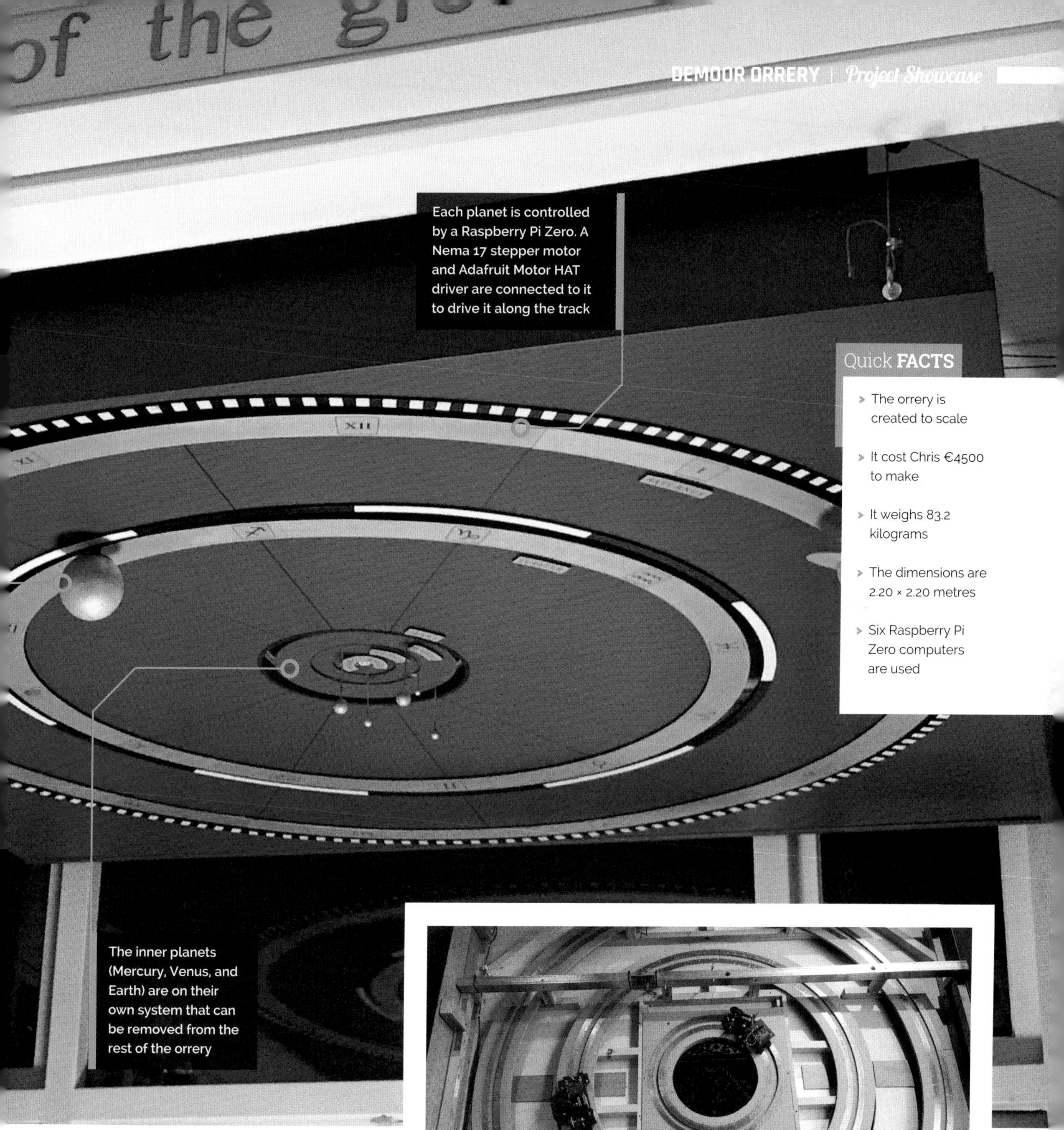

Each planet is controlled by a Raspberry Pi Zero. A Nema 17 stepper motor and Adafruit Motor HAT driver are connected to it to drive it along the track

The inner planets (Mercury, Venus, and Earth) are on their own system that can be removed from the rest of the orrery

Quick **FACTS**

- The orrery is created to scale
- It cost Chris €4500 to make
- It weighs 83.2 kilograms
- The dimensions are 2.20 × 2.20 metres
- Six Raspberry Pi Zero computers are used

Raspberry Pi in the sky

Even so, he persevered and learned, ending up with two ways of moving the planets, each of which is connected to a Raspberry Pi Zero computer. Mars, Jupiter, and Saturn are attached to front-wheel drive, 3D-printed cars which run on tracks on the non-visible side of the project. The inner planets, Mercury, Venus, and Earth, are mounted to dishes. "The inner system is so small, you can't have tracks there, so I mounted those planets on dishes and even connected Mercury directly to the axle of the stepper motor."

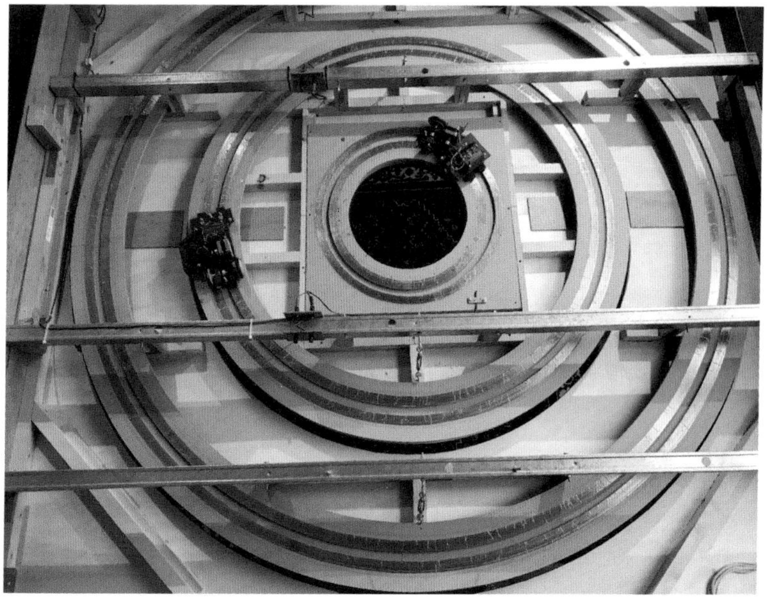

▲ To discover its position, each Raspberry Pi Zero computer detects when it passes a magnet placed on the track. This sets its counter to zero so it can work out how many steps the stepper motor must take to get a planet into position

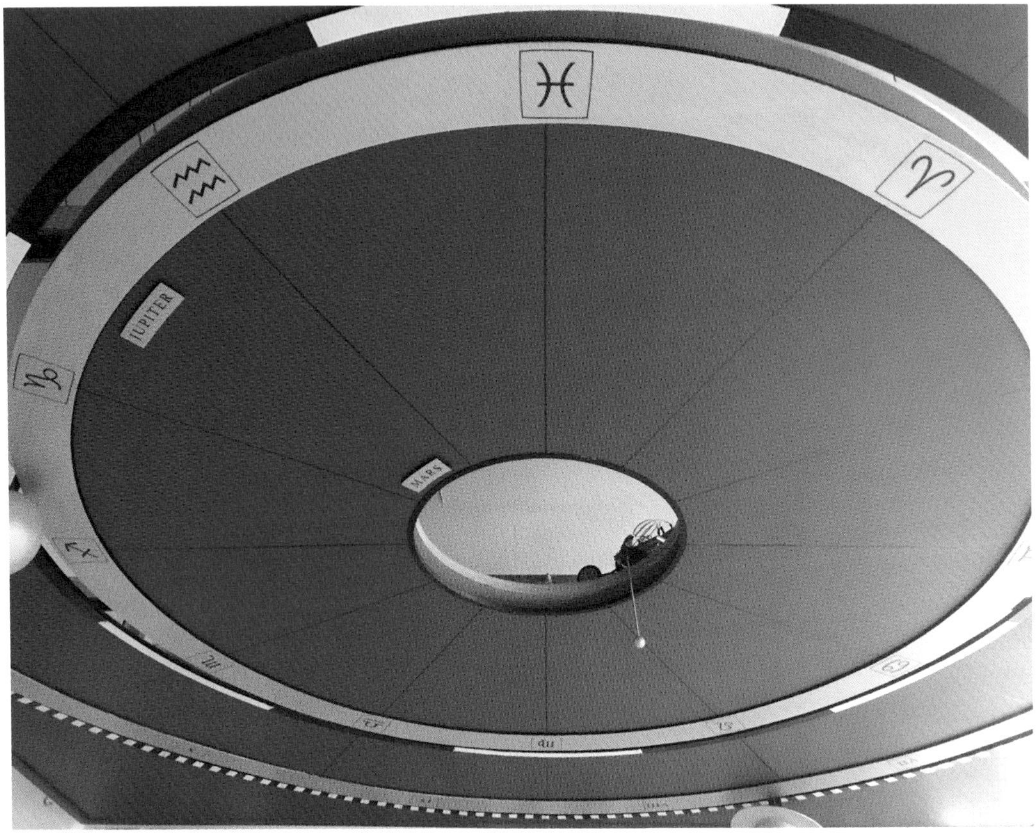

◀ The orrery is 30 centimetres high and suspended from the ceiling. Chris has attached it to a winch which allows it to be lowered for easy maintenance. A winding mechanism from a vacuum cleaner is used to ensure the power cord doesn't get in the way

▼ This car drives Jupiter around its circular track (in reality, orbits are not perfect circles). It gets electricity via two sliding contacts mounted on the back

▲ Here's a close-up view of the front of the Mars car with a Raspberry Pi Zero mounted on top

The driver runs on 12 volts and the Raspberry Pi Zero computers on 5 volts. The power supply is firmly attached to the frame

Warning!
Electrical Safety

Please be careful when working with electrical projects around the home. Especially if they involve mains electricity.

magpi.cc/electricalsafety

This is a view of the inner planets attached to their propeller dish. The Sun is drawn on the ceiling, otherwise it'd be 2.85 metres in diameter!

The biggest headache was working out the planetary positions. "I looked into the mathematics of NASA, and the exact positioning of Mercury is a mathematical equation, with more than 40 pages of data. It's very, very complex," Chris says. "I had to make it more simple, and I found a beautiful JavaScript library called JSOrrery which is installed on a server. You give it a date and it plots the planets. Raspberry Pi computers are then connected by wireless LAN and they read their position before moving their connected planets to it."

It's not 100 percent accurate. "If you want to travel to Mars and you let yourself be guided by my orrery, then you're definitely going to miss it," Chris laughs. But as a working showpiece, it's stunning and, what's more, it has been made open-source so that anyone can try to make their own. "Maybe it will lead to something – perhaps people will contact me and it'll involve some travelling and meeting new people," he says. "If not, I've got a beautiful ceiling and that was the whole point."

Recreating Eise Eisinga's historic orrery

01 Eise Eisinga spent seven years creating his orrery in the 18th century. Bought by King William I of the Netherlands in 1818, it was very detailed and also kept track of the phases of the moon. Credit: Erik Zachte, Wikipedia

02 Chris de Moor opted for a simpler system because of the smaller space he had. The outer planets are connected to 3D-printed cars running on tracks, controlled by a Raspberry Pi Zero.

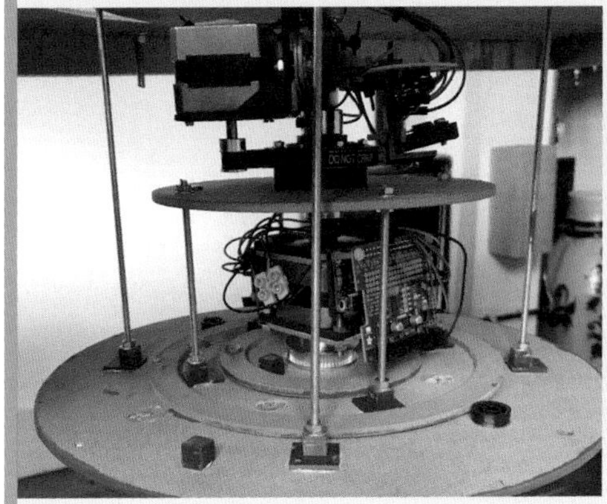

03 Inner planets Earth and Venus, meanwhile, are connected to a viewing dish that's mounted to a propeller dish. Mercury is mounted directly on to the stepper motor. It allows for tighter circling.

RMS meteor tracker

CCTV monitoring and Raspberry Pi enabled a stargazing duo to capture some incredible scenes, discovers **Rosie Hattersley**

MAKER

Mary and Mark McIntyre

Lifelong astronomy enthusiasts Mary and Mark McIntyre are part of the UKMON meteor network and enjoy sharing their astrophotography knowledge with others.

magpi.cc/marymcintyre

Mary and Mark McIntyre are so dedicated to astronomy that they had a mini-honeymoon at AstroFarm in France, and timed it so the moonlight wouldn't affect whether they could spot stars. They even moved halfway across the UK in order to set up home close to an important stargazing spot, at which point they decided to replace their existing meteor tracking setup. They have now built an RMS (Raspberry Pi Meteor Software)-based meteor tracker, one of dozens of astronomy trackers that helped pinpoint the whereabouts of a very rare carbonaceous chondrite meteorite which landed in Somerset in February 2020.

Mary and Mark's previous Raspberry Pi projects included a weather station and an all-sky camera which they used with an analogue meteor camera, and for which they adapted code to display findings on their website (see **github.com/markmac99**). Expense and maintenance issues with this setup meant they were only too happy to get involved with RMS Raspberry Pi meteor tracking instead. "The RMS project was conceived, and software is written collaboratively by astronomy academics who wanted a lost-cost DIY system using off-the-shelf parts, to generate science-grade data about meteors as they burn up in our atmosphere," says Mary.

This chimed with the interests of the McIntyres, who now run five meteor cameras and help co-ordinate the UKMON (UK Meteor Network). Its team of more than 100 UK citizen scientists has adapted RMS with a UK-specific toolset and data archive, making it far simpler for stargazers to identify events and objects they spot. "If more than one station captures the same event, then you can calculate velocity, mass, a more accurate orbit trajectory, and whether or not anything survived and fell as a meteorite," Mary explains. This is what happened with last year's Winchcombe fireball meteorite.

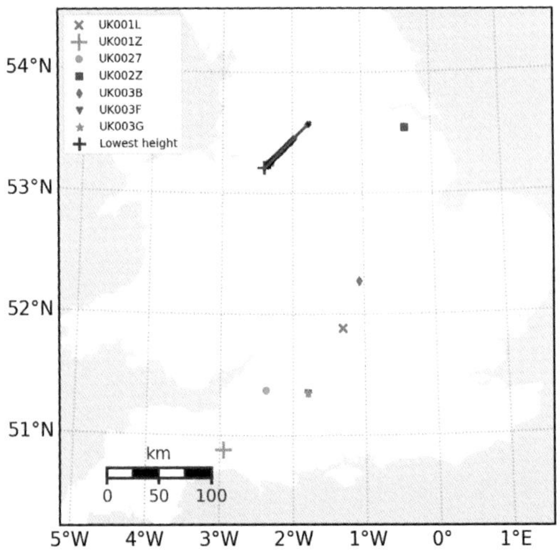

▲ Other cameras caught the same Perseid meteor and also the 'ground track' of the meteor as it travelled through the atmosphere and burned up

The meteor tracker once detected 20,000 objects an a single night (a spider had crawled over the lens)

Security cameras were chosen to track the night skies due to their reliable performance in low light

The camera is connected via Ethernet to a Raspberry Pi 3 with 128GB SD card to capture footage of passing meteors

Raspberry Pi stack running the four meteor tracking cameras

Quick FACTS

- RMS now has hundreds of members globally
- NASA uses RMS data to model impact hazards
- In November, a new meteor shower was identified
- The crucial data came from Mary and Mark's data
- It was the third meteor shower discovered by RMS

▲ A sun dog or parhelion, an atmospheric optical effect caused by ice crystals in high-level cirrus cloud, captured from the McIntyres' home observatory

> As well as meteorites, the RMS cameras can detect sprites – the extremely fast and faint upper atmosphere electric discharges that occur very high above thunderstorms

Scanning the skies

The McIntyres' meteor tracker uses a relatively inexpensive CCTV camera – chosen because it can detect objects in low light levels – which is pointing up at the sky and is clad in weatherproof housing. Handily, the camera can be positioned several metres from the couple's home, while its Raspberry Pi is tucked away indoors. Mary's version uses a Raspberry Pi 3, but she recommends the additional power of Raspberry Pi 4 if you're keen to build your own. "We sometimes detect several hundred meteors per night on every camera, which can lead to heavy load on Raspberry Pi."

Connectivity is in the form of PoE (Power over Ethernet), but there are other options: "The biggest advantage of Raspberry Pi is that the system can be installed virtually anywhere. Ideally, it should have an internet connection, but it can be run off-grid using batteries and solar power if necessary, provided the operator can

UK001L 2021-08-12 23:43:10 UTC

▲ Bright Perseid meteor – magnitude -2.8 (the brighter something is, the lower the number) caught on the McIntyres' camera UK001L

visit periodically to collect data," says Mary. Parts to build the meteor tracker were easy to source from The Pi Hut and other component suppliers, with a total cost of around £200, including the camera. "If you have a little experience using Raspberry Pi and Linux, it's pretty simple to set up," she says. There's a pre-built image suitable for UK users on their website (**magpi.cc/rmswiki**).

Wonders of the universe

The free open-source Raspberry Pi Meteor Software (RMS) records the sky continuously from dusk until dawn, looking for anything that is moving. RMS recognises and discounts aircraft and satellite trails, fireworks, and most moths, bats, and owls. The camera is calibrated in advance with a 'platepar' file that identifies the star field in its field of view. Based on this information, it is able to work out the orbit of the material that caused a meteor event.

As well as meteorites, the RMS cameras can detect sprites – the extremely fast and faint upper atmosphere electric discharges that occur very high above thunderstorms. This is helping researchers who study upper atmospheric phenomena.

The open-source software is primarily written in Python 3 (and supports Python 2) with some performance aspects coded in C/C++. Downloadable from **magpi.cc/rmsgit**, its video capture and stability have recently been improved, while machine learning and detection accuracy are ongoing.

Mary has been impressed at how well Raspberry Pi has worked for their meteor detection setup. It cost around a third as much as a previous meteor system based on Windows. "Raspberry Pi has made it much more affordable, much easier to set up, and much simpler to maintain."

Enable night mode

The RMS software can be downloaded from **magpi.cc/rmswiki**, while detailed build instructions are at **magpi.cc/rmscamera**. You'll need a Raspberry Pi 3 or 4, IP camera or Raspberry Pi HQ Camera, large-capacity microSD card, and an Ethernet connection.

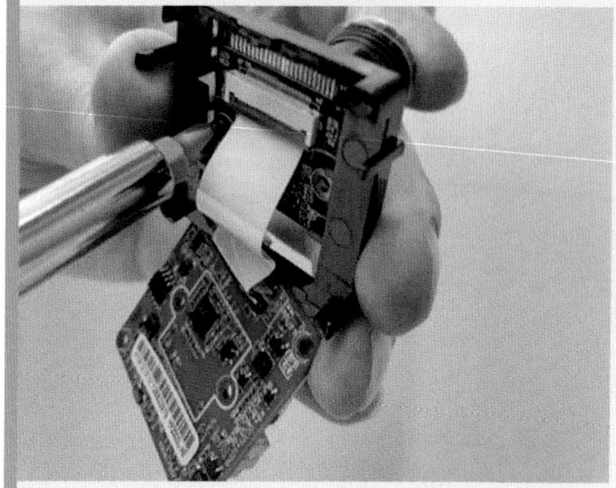

01 Connect the camera via Raspberry Pi's Ethernet port and a PoE injector, and run the script on Raspberry Pi to configure the camera for night vision.

02 Install the camera in a weatherproof housing on the outside of a building with a good view of the night sky.

03 After the first night's run, 'calibrate' the camera's field of view against the stars – see the setup guide. After this, it should be set to run automatically every night.

Teasmade 2.0

Here's a fresh hack on an absolute classic. Hot beverage anyone? **Nicola King** admits to being old enough to remember the original

MAKER

Martin Spendiff and Vanessa Bradley

Martin is a mathematical modeller who left the UK for Switzerland. He's a fan of FOSS and tech that serves users. Vanessa is new to coding, and a constant source of weird/good ideas.

magpi.cc/teasmade2

Warning!
Mains electricity

Martin added a relay switch to the innards of the machine. That involves wires that carry mains voltage. Anything with relay switches needs to be done with care.
magpi.cc/electricalsafety

Teasmades are machines for automatically making tea. They were hugely popular in the 1960s and 1970s, and often placed in bedrooms (for the convenience of having a piping hot cup as soon as you were awoken by the machine's alarm). The Goblin Teasmade, in particular, is something of an icon in its field.

This author fondly remembers her late grandmother's Teasmade that sat proudly in a guest bedroom, waiting for the odd bed-and-breakfast guest that she occasionally took into her home. Although, we did always wonder the point of it when you still had to get up and go to the kitchen to get some milk from the fridge. But, despite our puzzlement, the Teasmade has basked in a form of legendary glory ever since.

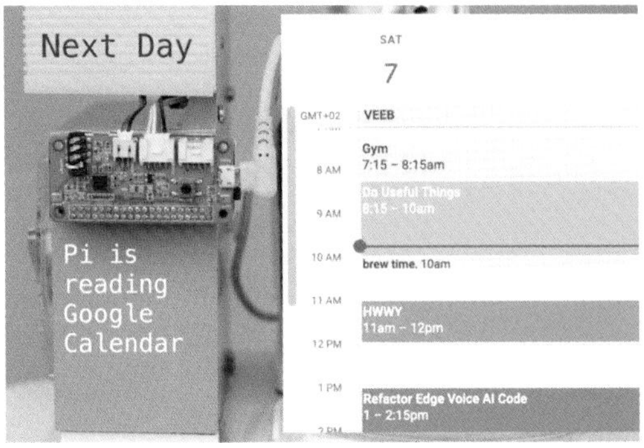

▲ Raspberry Pi receives alerts from Google Calendar so it knows when to start making a brew

▼ The innards of the Teasmade, fitted with a new Grove Two-Channel Solid-State Relay

▲ A speaker was added inside the light fitting on top of the Teasmade

The coffee stimulus

Hardly surprising then that, decades later, some makers are keen to revive that concept of an automatic hot drink maker, and what better base to use for it than a retro Goblin Teasmade? Martin Spendiff and Vanessa Bradley are two such makers, and this fun update uses a Teasmade along with Raspberry Pi Zero WH to produce their hot drink of choice... coffee.

Their motivation for the project was twofold, as Martin explains: "The Teasmade (inherited) is pretty impressive – it has no microprocessors and therefore has a workflow built with just a few switches and logic. I've long admired the design, so bringing the innards up to date by adding a little Linux machine to the mix was appealing. I don't like to say cyberpunk, but cyberpunk."

He also wanted to incorporate Google Calendar into the venture. "It's nice to have a serious idea dressed up in a novelty application," he tells us. "Getting Google Calendar entries to trigger events has crossed my mind as a useful idea a few times. The idea of automating a cup of coffee was enough motivation to get it over the finish line."

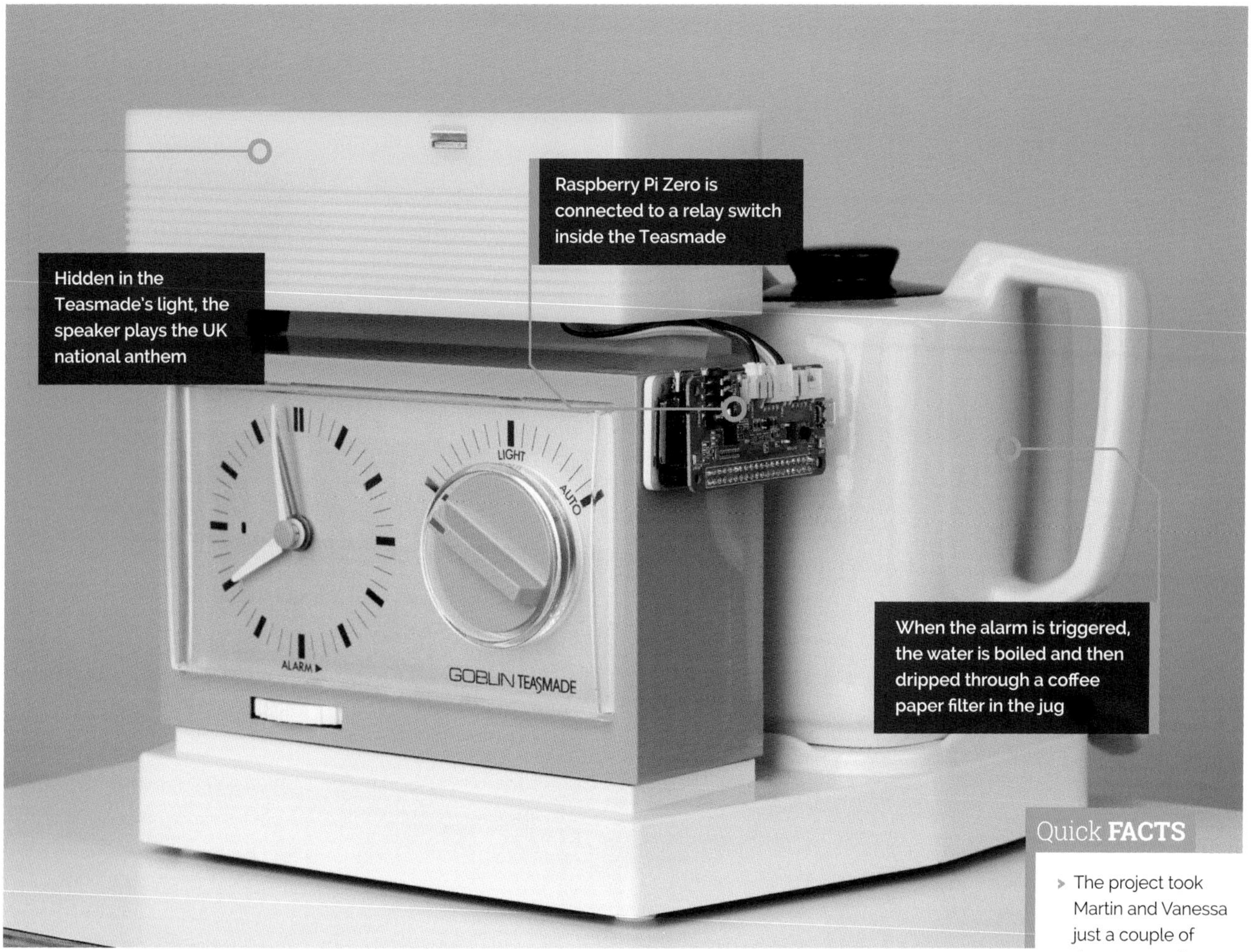
Raspberry Pi Zero is connected to a relay switch inside the Teasmade
Hidden in the Teasmade's light, the speaker plays the UK national anthem
When the alarm is triggered, the water is boiled and then dripped through a coffee paper filter in the jug
LIGHT
AUTO
ALARM
GOBLIN TEASMADE

Right royal cuppa

So, how exactly does the machine produce a regular caffeine hit? On the side of the Teasmade is a Raspberry Pi Zero with a Grove ReSpeaker HAT. "There is also a connection to a little cheap speaker and a relay switch, replacing the alarm switch on the inside," says Martin. "The script is automatically started using systemd when it turns on. It starts a process that monitors a Google Calendar. Every minute, it runs a query on the calendar, looking for the trigger phrase. If it sees it, it starts the boil cycle."

The most difficult challenge for the pair was working out how to get the tool that pulls alerts from a Google Calendar (gcalcli) to act as a trigger in the main code. In the end, they used the subprocess module in Python, and now the Teasmade 2.0 cleverly starts brewing their coffee ten minutes before the time it's required.

What's more, the speaker that Martin and Vanessa inserted belts out the *God Save the Queen* while refreshments are being prepared. "There is a recurring appointment in our Google Calendar for an 11 am coffee. The sound of the tinny national anthem wafting from the kitchen is our signal that it is time to down tools for a break," says Martin. This has generated some "bewildered amusement" from their non-British pals: "Friends and family are baffled by it on a number of levels."

> "The sound of the tinny national anthem wafting from the kitchen is our signal that it is time to down tools for a break"

If you'd like to create your own version, Martin has generously made the code for the project available on GitHub, and one motivation for this is the old adage, 'the more the merrier'. "Generally, the more people that replicate it and chip in, the easier it gets," notes Martin. "Better coders than me have refined my code on other projects – I've no doubt that the same would apply here."

Quick FACTS

- The project took Martin and Vanessa just a couple of days to complete
- The LEDs flash a patriotic red, white, and blue while the drink is prepared
- Find the code and instructions on GitHub: **magpi.cc/teasmadegit**
- Check out their other great makes here: **magpi.cc/veebyoutube**
- Martin recommends YouTuber James Hoffmann's musings on bedside coffee makers: **magpi.cc/barisieur**

Maker Guides

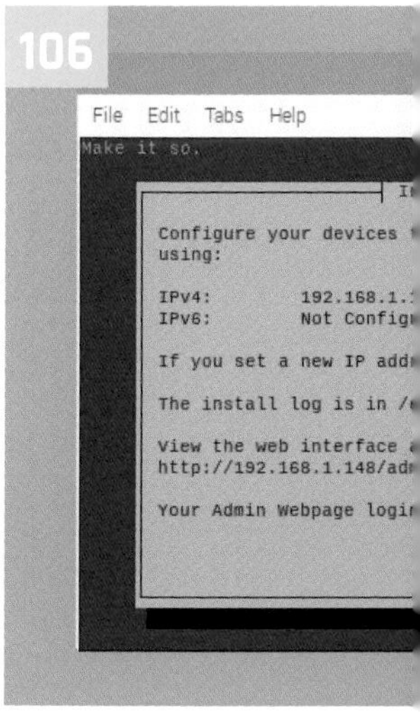

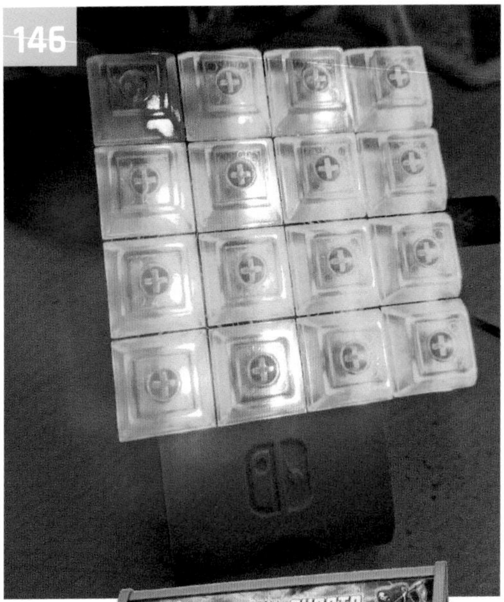
146

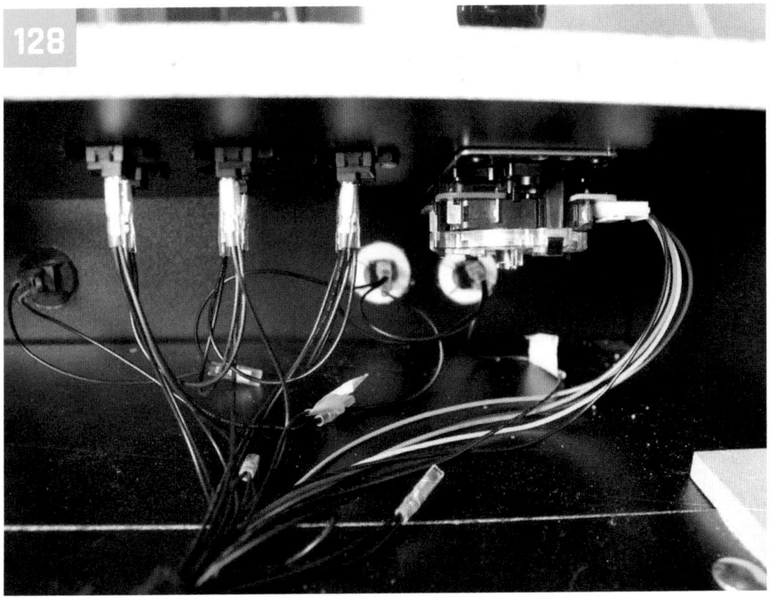
128

140
GHOULS'N GHOSTS

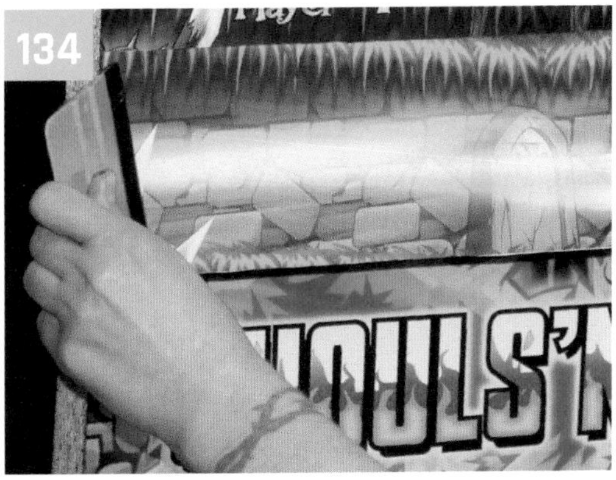
134

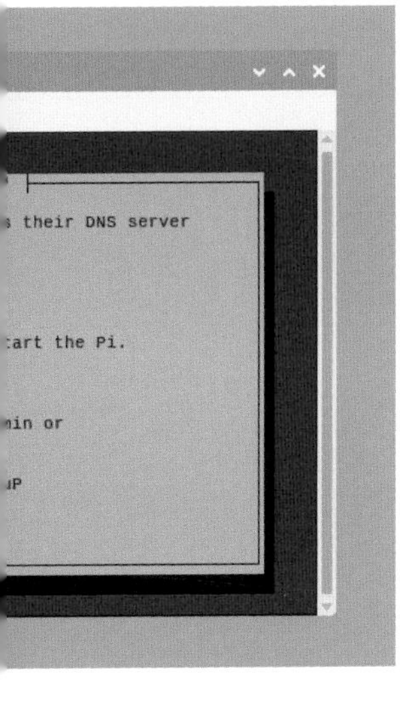
s their DNS server
tart the Pi.
min or

Set up Pi-hole with Raspberry Pi

Pi-hole blocks ads, cookies – and whole sites if you choose – from your home network so you can browse more quickly and securely

MAKER

Nik Rawlinson

Esperanto-speaking, pencil-wielding, single-board computing fan who likes hyphens and remembers what that icon on the save button depicts.

nikrawlinson.com

Pi-hole is a free web filtering tool that runs on a Raspberry Pi on your network. By connecting your other computers, tablets – and even your smartphone – to Raspberry Pi, rather than directly to your router, Pi-hole will interrogate their internet traffic and strip out unwanted content. With built-in lists for ad servers, its primary use is to block advertising, but you can just as easily bar social media and other sites you find distracting or objectionable. In this tutorial, we'll show you how to set up Pi-hole on a Raspberry Pi and connect to it from another device on your network.

01 Locate Raspberry Pi

Start with a fresh installation of Raspberry Pi OS connected to your local network (via an Ethernet cable or wireless LAN). Pi-hole only works if the other computers on your network know where to find it. Most routers assign IP (internet protocol) addresses dynamically, and there's a chance your Raspberry Pi might move around the network and get a new IP address. To ensure your other devices can always find it, we're going to give your Raspberry Pi a static IP address. Open a Terminal window by clicking the icon on the Raspberry Pi menu bar and enter:

```
hostname -I
```

This will tell you which IP address is currently assigned to your Raspberry Pi. Ours is 192.168.1.148. Make a note of this number.

▼ Make sure Pi-hole is set to filter content on the interface through which it's connected to your network

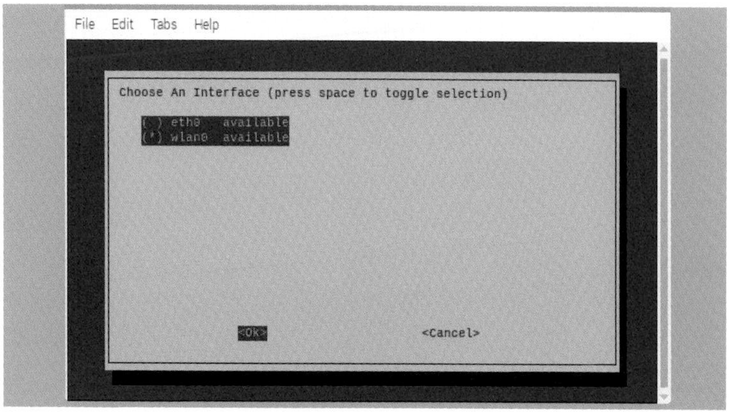

02 Get your router's IP address

Now do the same for your router's IP address (the 'default gateway'). Enter this command in Terminal:

```
ip r
```

Press **ENTER**, and note the first four sets of digits after 'default via'. Ours is '192.168.1.1'. This is your router's address.

03 Update your configuration file

We're going to add this info to our configuration file so it never changes. Open Terminal and enter:

```
sudo nano /etc/dhcpcd.conf
```

At the bottom of the **dhcpcd.conf** file you will see an example of a static IP configuration. Delete the hashtags ('#') at the start to uncomment the 'static IP address' and 'static IP routers' lines. Replace the example IP address with the number from Step 1 and your router address from Step 2. See the **dhcpcd.conf**

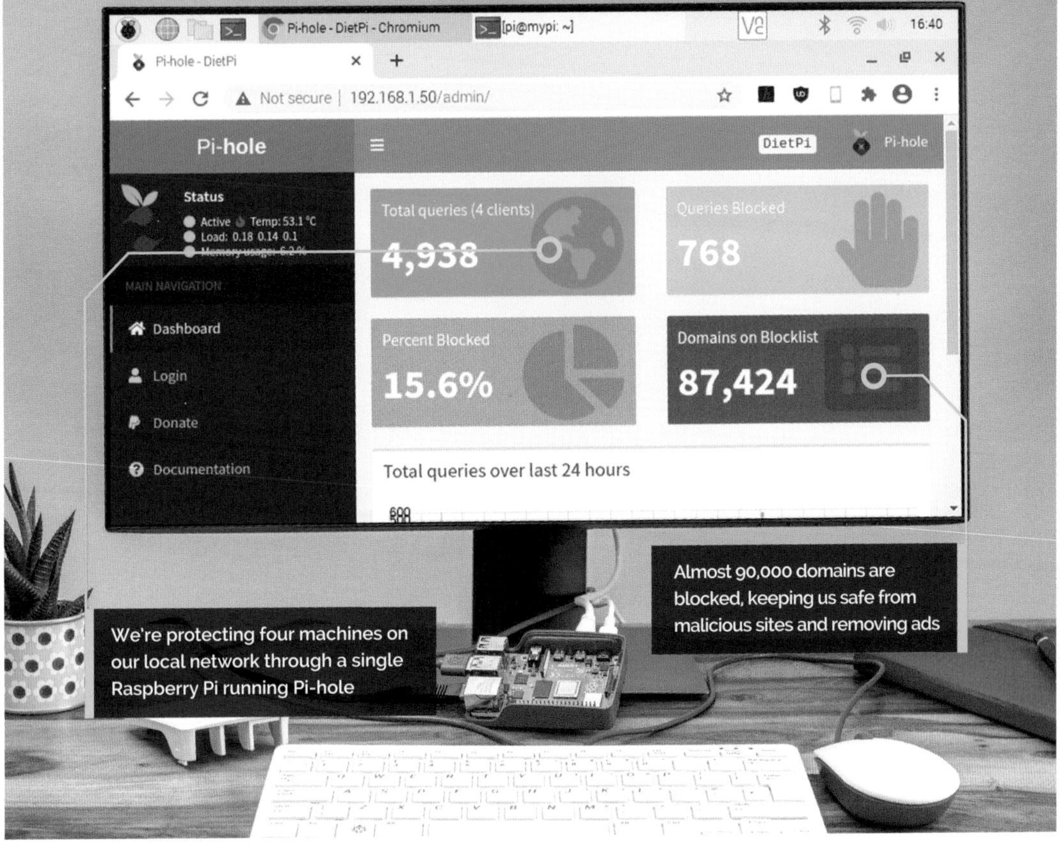
Pi-hole
Status
Total queries (4 clients)
4,938
Queries Blocked
768
Percent Blocked
15.6%
Domains on Blocklist
87,424
Dashboard
Login
Donate
Documentation
Total queries over last 24 hours
We're protecting four machines on our local network through a single Raspberry Pi running Pi-hole
Almost 90,000 domains are blocked, keeping us safe from malicious sites and removing ads

code listing for an example of how our configuration looks (but don't forget to use your own IP and router addresses).

If your Raspberry Pi is connected to the router via an Ethernet cable, remove the comment '#' before `interface eth0`. If you are connecting Raspberry Pi to your network with wireless LAN, replace it with `interface wlan0`.

Press **CTRL+O** to save your file ('O' stands for output) and **CTRL+X** to quit Nano, then enter `sudo reboot` and press **ENTER** to restart.

04 Download the Pi-hole installer

When it's up and running again, you're ready to install Pi-hole. First, download a copy of the latest build. Open a new Terminal window and enter:

```
wget -O basic-install.sh
https://install.pi-hole.net
sudo bash ./basic-install.sh
```

This will start the Pi-hole installation script. It will check which packages are already installed, and install the ones you need. Follow the instructions in the installation while reading through the following steps.

05 Specify your interface

After some preliminary configuration, the splash screen for the Pi-hole automated installer will appear, explaining that it's about to "transform your device into a network-wide ad blocker". Press **ENTER** on each of the first three screens. On the 'Choose an interface' screen, use the arrows and **SPACE** bar to select either eth0 or wlan0 for a wired or wireless connection, as appropriate to your setup. Press **TAB** to select OK and hit **ENTER**.

> Pi-hole uses an external DNS provider to locate authorised resources

06 Choose a DNS provider

Any internet resources, like images, text and code, that don't trip Pi-hole's safeguards will be retrieved from the servers where they're stored and fed back to the device that requested them. Pi-hole uses an external DNS provider to locate these authorised resources, and gives you a choice of nine to pick from, plus a custom option for business users running their own DNS server. If you're happy to stick with the default, which is Google, press **TAB** to select OK, then press **ENTER**. If not, select an alternative from the list, press **TAB** to select OK, then **ENTER** to move to the next page.

07 Confirm your block list

To save you specifying every server that should be blocked, Pi-hole is configured to use a pre-compiled list ('StevenBlack') to which you can add your own entries once it's up and running. Leave the

Top Tip

Change password

You can change your password at the command prompt with the command: `sudo pihole -a -p`

You'll Need

- Raspberry Pi
- Pi-hole installer **pi-hole.net**
- Your router details

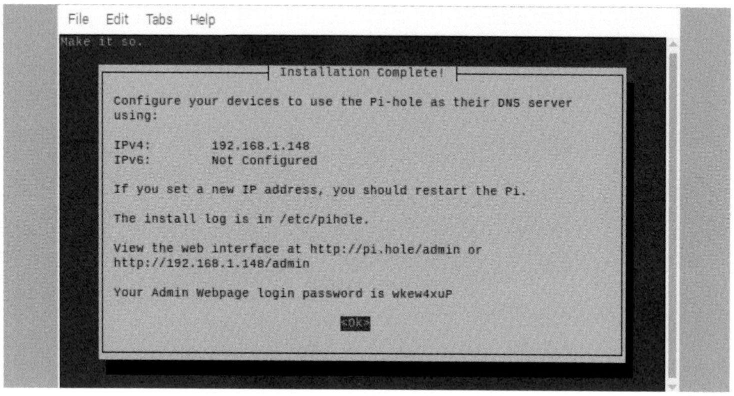
File Edit Tabs Help
Make it so.
Installation Complete!
Configure your devices to use the Pi-hole as their DNS server using:
IPv4: 192.168.1.148
IPv6: Not Configured
If you set a new IP address, you should restart the Pi.
The install log is in /etc/pihole.
View the web interface at http://pi.hole/admin or http://192.168.1.148/admin
Your Admin Webpage login password is wkew4xuP
<Ok>

▲ When you've finished setting up Pi-hole, make a note of the IPv4 address and password for later use

existing list selected by pressing **TAB** to select OK and press **ENTER**. Ensure both IPv4 and IPv6 are selected on the 'Select Protocols' screen, and select OK.

At this point, the installer checks that you're happy with the numeric (IP) address Raspberry Pi is using. Check that the IP address and Gateway match your IP address and router's address. As we configured this in Step 3, and set it to remain fixed, step through the next two screens without worrying too much over the warnings that static addresses might cause conflicts.

08 Install the web interface

You can use Pi-hole's web interface to monitor your web traffic and temporarily deactivate web filtering when you need to access resources that would otherwise have been blocked. It's also where you'll add entries to and remove them from the block list. Make sure 'On' is selected when asked if you want to install the web admin interface, then press **TAB** to select OK and press **ENTER**. Do the same on any following screens to install the lighttpd web server that will host the web interface, and the PHP modules that it relies on.

09 To log or not to log

When you reach the 'log queries' screen, decide whether you want to log queries or not. They're useful if you want to look back and see what's been requested and blocked on your network, but not critical to the effective running of Pi-hole. If you want to save space on your Raspberry Pi's microSD card, or reduce the number of times your Raspberry Pi writes to it (media cards can sustain a generous but limited number of write and wipe operations before they start to fail), you can turn off this option. We're going to leave it on.

10 Select a privacy mode

How much do you want to know about what's happening on your network? Pi-hole next asks what level of detail it should record in its stats. Level 0, the default, logs everything, with higher numbers anonymising more and more data. If you're never going to need to know what other users on your network are up to, you can set it to 3, but sticking with level 0 can be useful if you suspect processes installed somewhere on your network might be sending your personal information to remote servers, as you can trawl the logs for unfamiliar hosts.

11 Specify your interface

Pi-hole has all the information it needs to complete the setup process, the remainder of which is automated.

When it's completed, take a note of the IPv4 address displayed on the final screen, and the Admin login password. Quit the installer and reboot Raspberry Pi with:

```
sudo reboot
```

At this point you might want to remove the screen, keyboard, and mouse. Locate your Raspberry Pi near your router if using wireless LAN or keep it connected via the Ethernet cable. From now on, we will access it via the local network.

12 Using Pi-hole

Now switch to another computer on your network. It can be another Raspberry Pi, or a Mac or PC (or even a smartphone or tablet). You will need to change the network settings on every device you want to filter through Pi-hole so it accesses your Raspberry Pi rather than your router directly.

Top Tip

Permanently powered

Keep the Raspberry Pi running Pi-hole switched on. If not, your connected devices will resort to the fallback DNS server.

dhcpcd.conf

> Language: **Bash**

DOWNLOAD THE FULL CODE: magpi.cc/github

```
interface wlan0 # use eth0 for wired Ethernet
static ip_address=192.168.1.148/24 # Use your Raspberry Pi's IP address
static routers=192.168.1.254 # Use your Router's IP address
static domain_name_servers=192.168.1.254 8.8.8.8
```

Manually configure each device for Pi-hole

Configure a Windows device

On a Windows computer, open Control Panel and click 'View network status and tasks'. Click the hyperlink of your network connection (Ethernet or WiFi, followed by Properties. Double-click 'Internet Protocol Version 4 (TCP/IPv4)'. Click the radio button beside 'Use the following DNS server address' and, in the box beside 'Preferred DNS server', enter the IP address of your Raspberry Pi running Pi-hole.

In the 'Alternative DNS server' box, enter either your router address, if you want to fall back on your ISP's DNS server should there be any problem with Pi-hole, or 8.8.8.8 if you want to use Google's DNS server instead.

Make sure Pi-hole is set to filter content on the interface through which it's connected to your network

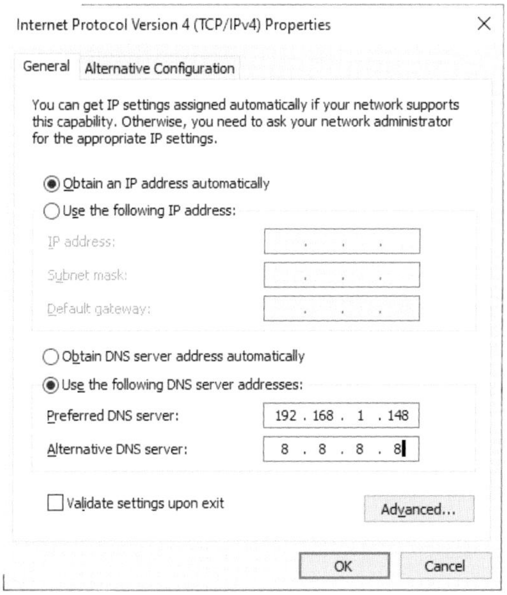

Set up macOS

Open System Preferences and choose Network. Choose the network interface from the sidebar (typically Wi-Fi or Ethernet) and click 'Advanced'. Choose the DNS tab and click the '+' (Add a DNS Server icon). Type the address of your Raspberry Pi running Pi-hole in the line that appears. You can optionally add another server, like your router's IP address or 8.8.8.8 to use Google's DNS server

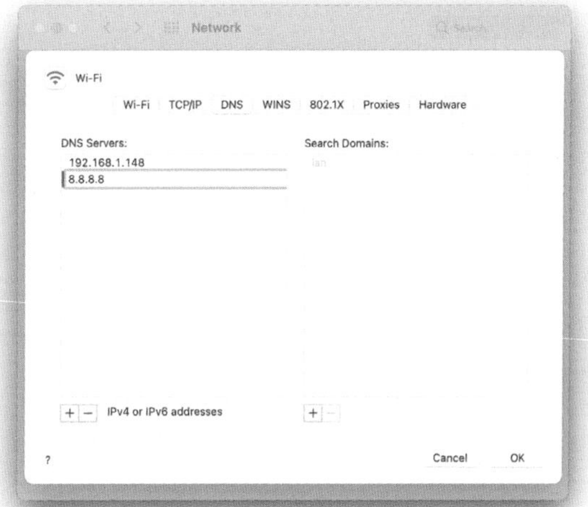

Specifying a second DNS address gives your computer a fallback that it will use if it can't reach your Pi-hole device

as a fallback in the case of problems. Select the addresses of any existing DNS servers in the left-hand box, and click '–' to delete them.

Set up Linux

To set up another Raspberry Pi to use Pi-hole as its DNS server, switch to that Raspberry Pi, open Terminal and type:

```
sudo nano /etc/dhcpcd.conf
```

Move your cursor to the bottom of the file and add the following, replacing `[ip address]` with the numeric address of the device running Pi-hole:

```
static domain_name_servers=[ip address]
```

Press **CTRL+X** to exit and confirm that you want to save the file when asked. Now restart the dhcpcd service by typing:

```
sudo service dhcpcd restart
```

On a Linux PC running the Gnome interface, launch the Settings app and click Wi-Fi in the sidebar, followed by the cog beside your active network name. Click the IPv4 tab, then turn off the switch beside Automatic to the right of DNS. Enter the numeric address of your Pi-hole device in the field below, then click Apply.

Top Tip

Beware changing networks

When you move your computer to a different network, you'll bypass Pi-hole. Change your DNS settings for every network you use.

Run Pi-hole on your Raspberry Pi and enjoy the worldwide web without all those annoying ads

Configure Pi-hole's DNS settings

Beef up your privacy and give computers on your network their own web addresses by tweaking Pi-hole's DNS settings

MAKER

Nik Rawlinson

Esperanto-speaking, pencil-wielding, single-board computing fan who likes hyphens and remembers what that icon on the save button depicts.

nikrawlinson.com

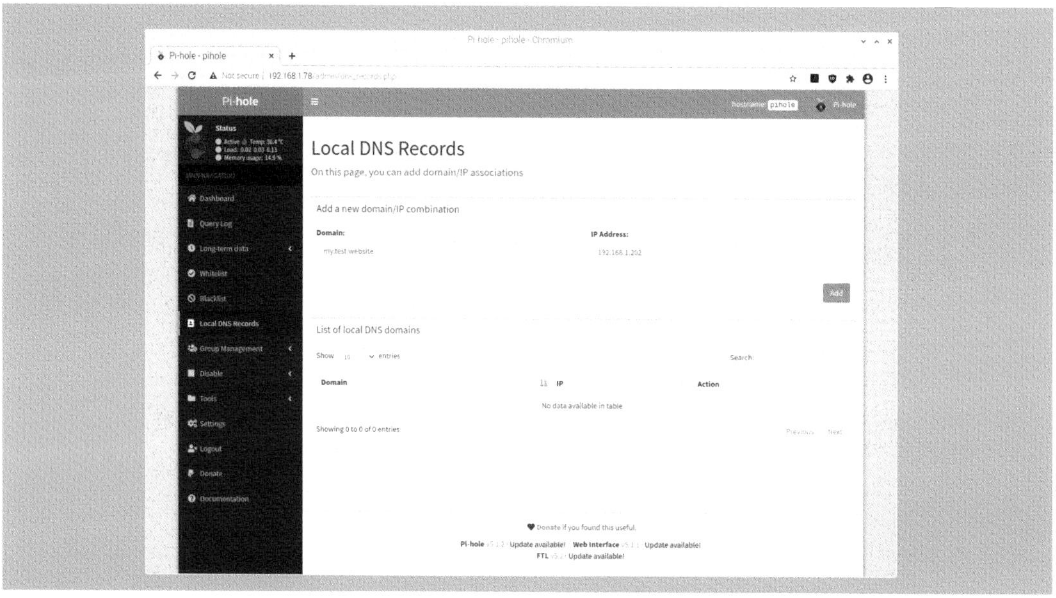

If you're hosting or developing a website on your local network, Pi-hole lets you assign a memorable name in place of its IP address

In the previous tutorial, we set up Pi-hole to filter web content. The result was less cluttered web pages, free of ads, that loaded more quickly. Here, we'll take things further by setting up a local DNS (Domain Name System) server so Pi-hole can bypass third-party DNS providers to find a direct route to the web resources we need, for greater security. We'll also set up some friendly web addresses for computers on our local network. That way, if we're using them for tasks like building and testing websites, we can access them directly through a browser without having to use their numeric IP addresses.

01 Log in to Pi-hole

Open a new browser window and log in to your Raspberry Pi that's hosting Pi-hole. If you're using a computer that's configured to use Pi-hole to filter its content, you'll find it by entering:

http://pi.hole/admin

If not, you'll need to use the Pi-hole Raspberry Pi's IP address, which you set up during installation. For example:

http://192.168.1.148/admin

There's no username, but you will need your Pi-hole password. Click Login and enter it.

02 Creating local DNS records

You'll notice that one of the options for logging into Pi-hole was to type **pi.hole** into the

You'll Need

- Pi-hole **pi-hole.net**
- Web browser
- Optionally: Terminal access

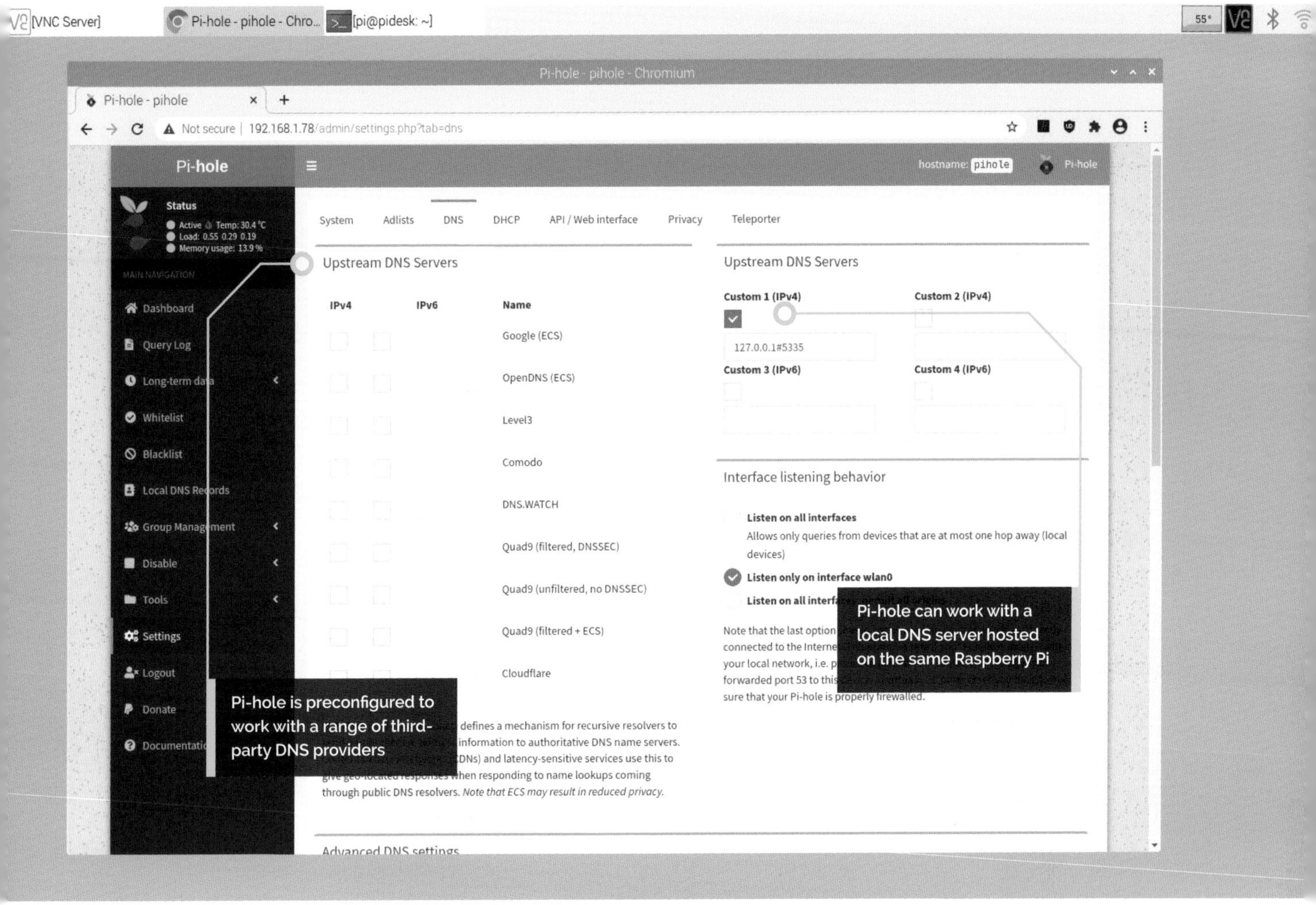

browser, rather than use the Raspberry Pi's IP address. This works because when your computer passes the request to Pi-hole, it first looks for a matching local record before passing it on to an external DNS server. Finding a match, it checks the address of the host that it points to and delivers whatever result it finds there. You can apply the same trick to give web addresses to any computer on your network.

> Choose a memorable address and type it into the Domain box

03 Friendly names, locally

If you're running a web server on one of your computers, make a note of its IP address. Now click Local DNS Records in the Pi-hole sidebar. Type the numeric address of the computer hosting your development server in the IP Address box in the 'Add a new domain/IP combination' panel. Choose a memorable address and type it into the Domain box; for example, 'my.test.site'. Click Add and, within a couple of seconds, the DNS table will be updated. Open a browser window on any device filtered by Pi-hole and type the new address, prefixed by **http://**.

04 Switch DNS server

When setting up Pi-hole, you selected an external DNS provider, which is used to look up the addresses of permitted web-hosted assets. If you want to change this, click Settings in the Pi-hole sidebar, followed by the DNS tab. You'll notice that at least two ticks appear alongside your preferred service (four if you're filtering both IPv4 and IPv6 requests), for the primary and secondary servers. If you choose to use a secondary provider, either swap both ticks, as appropriate, or mix and match. You could, for example, use Google and Cloudflare.

Top Tip

Get started

The first part of this tutorial series appeared in *The MagPi* magazine issue 104
magpi.cc/104

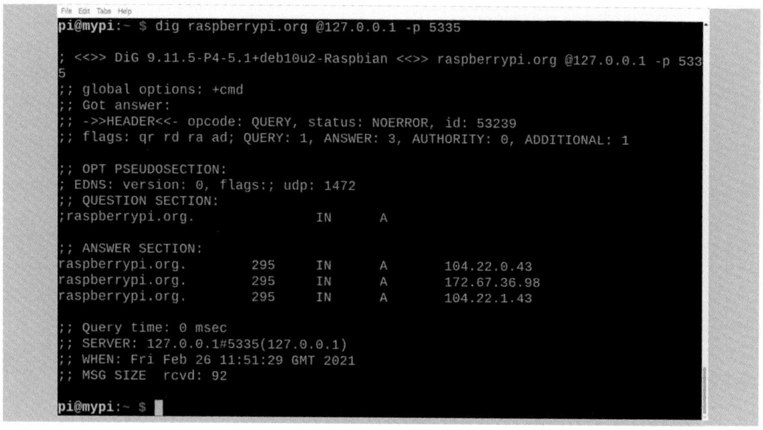
File Edit Tabs Help
pi@mypi:~ $ dig raspberrypi.org @127.0.0.1 -p 5335
; <<>> DiG 9.11.5-P4-5.1+deb10u2-Raspbian <<>> raspberrypi.org @127.0.0.1 -p 5335
;; global options: +cmd
;; Got answer:
;; ->>HEADER<<- opcode: QUERY, status: NOERROR, id: 53239
;; flags: qr rd ra ad; QUERY: 1, ANSWER: 3, AUTHORITY: 0, ADDITIONAL: 1
;; OPT PSEUDOSECTION:
; EDNS: version: 0, flags:; udp: 1472
;; QUESTION SECTION:
;raspberrypi.org. IN A
;; ANSWER SECTION:
raspberrypi.org. 295 IN A 104.22.0.43
raspberrypi.org. 295 IN A 172.67.36.98
raspberrypi.org. 295 IN A 104.22.1.43
;; Query time: 0 msec
;; SERVER: 127.0.0.1#5335(127.0.0.1)
;; WHEN: Fri Feb 26 11:51:29 GMT 2021
;; MSG SIZE rcvd: 92
pi@mypi:~ $

When you've set up Unbound, use the dig command in the Terminal to check that it can locate online resources

If you choose an alternative provider, it will use whichever DNS responds first (see **magpi.cc/roundrobindns** for more information).

Top Tip

Reliable connections

Consider using Ethernet, rather than WiFi, if your Raspberry Pi has it, to avoid DNS errors on your network caused by weak wireless signals.

05 Horses for courses

Different DNS services support different features. Cloudflare, for example, makes a point of not logging the IP addresses of browsers using its DNS server, and both OpenDNS and Comodo maintain lists of phishing sites, which should be blocked in addition to anything on your Pi-hole blacklist. OpenDNS Family Shield, which doesn't appear in the list of presets, not only blocks malware, but pornography too. To enable it, tick the checkboxes beneath 'Custom 1 (IPv4)' and 'Custom 2 (IPv4)' in the right-hand box and enter the addresses **208.67.222.222** and **208.67.220.220** in the input boxes. Scroll down and click Save.

06 A local DNS server

Alternatively, you may choose to run your own local DNS server. Why? Because more security-conscious system admins can be wary of trusting third-party providers rather than referring back to the most definitive sources for every address the computers on their network request. Pi-hole makes this possible by piggybacking the open-source Unbound server app. This isn't installed by default as part of Raspberry Pi OS, so open a Terminal window on the device hosting Pi-hole and type `sudo apt install unbound`. When asked to confirm the installation, press **Y**.

07 Configure Unbound

When the installation has completed, type `sudo nano /etc/unbound/unbound.conf.d/pi-hole.conf` and press **ENTER**. Pi-hole's developers have helpfully provided a complete configuration file at **magpi.cc/piholeunbound**. Copy everything in the large grey box in the Configure Unbound section, then press **CTRL+X** to quit Nano. Confirm that you want to save the configuration file when asked, and accept the suggested file name. Now restart the DNS server by typing `sudo service unbound restart` and pressing **ENTER**. Your Raspberry Pi is now running its own DNS server that interrogates the internet's root servers, rather than preconfigured DNS servers, to locate the web resources you need.

08 Test your DNS server

Test Unbound by typing `dig raspberrypi.org @127.0.0.1 -p 5335`. This invokes the Linux dig tool to query the DNS server hosted on port 5335 at the IP address 127.0.0.1 for the record that relates to **raspberrypi.org**. The address 127.0.0.1 effectively loops whatever is pointed at it back on itself, whether it's a browser request or, in this case, a command. Within a second, Unbound will have found the servers hosting the Raspberry Pi website and returned their IP addresses.

09 Check DNS caching

Once Unbound has identified the numeric addresses that relate to the web address you fed it, it saves them to its local database so that the next time the same resource is requested it doesn't need to repeat the operation. Press the up arrow key to reload the command you just typed and press

Pi-hole lets you choose your preferred DNS provider during setup, but you can switch to an alternative through the Dashboard

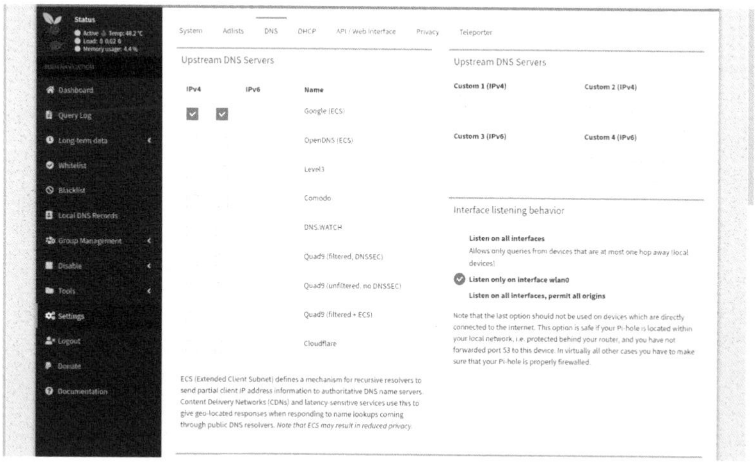
Status
Dashboard
Query Log
Long-term data
Whitelist
Blacklist
Local DNS Records
Group Management
Disable
Tools
Settings
Logout
Donate
Documentation
System
Adlists
DNS
DHCP
API / Web interface
Privacy
Teleporter
Upstream DNS Servers
IPv4
IPv6
Name
Google (ECS)
OpenDNS (ECS)
Level3
Comodo
DNS.WATCH
Quad9 (filtered, DNSSEC)
Quad9 (unfiltered, no DNSSEC)
Quad9 (filtered + ECS)
Cloudflare
Custom 1 (IPv4)
Custom 2 (IPv4)
Custom 3 (IPv6)
Custom 4 (IPv6)
Interface listening behavior
Listen on all interfaces
Listen only on interface wlan0
Listen on all interfaces, permit all origins

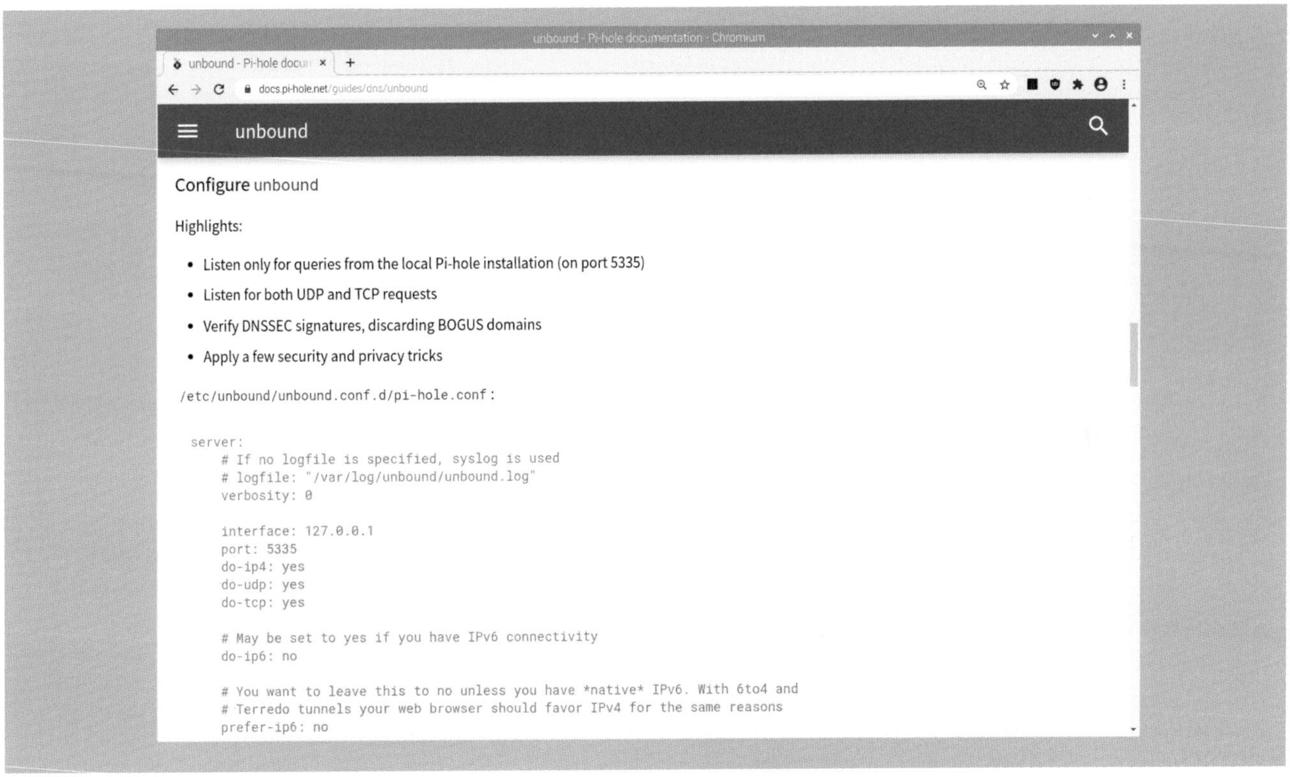
unbound - Pi-hole documentation - Chromium
unbound - Pi-hole docu
docs.pi-hole.net/guides/dns/unbound
unbound
Configure unbound
Highlights:
Listen only for queries from the local Pi-hole installation (on port 5335)
Listen for both UDP and TCP requests
Verify DNSSEC signatures, discarding BOGUS domains
Apply a few security and privacy tricks
/etc/unbound/unbound.conf.d/pi-hole.conf:
server:
If no logfile is specified, syslog is used
logfile: "/var/log/unbound/unbound.log"
verbosity: 0
interface: 127.0.0.1
port: 5335
do-ip4: yes
do-udp: yes
do-tcp: yes
May be set to yes if you have IPv6 connectivity
do-ip6: no
You want to leave this to no unless you have *native* IPv6. With 6to4 and
Terredo tunnels your web browser should favor IPv4 for the same reasons
prefer-ip6: no

Pi-hole's help pages include a comprehensive configuration file for using Unbound, which you can paste directly into your own config files

ENTER, and you may notice that the response is slightly faster the second time around. That's because Unbound is what's known as a caching server: it caches its work for future reuse. You now need to tell Pi-hole to use this instead of one of its preset DNS options.

10 Enable your DNS server

Return to the Pi-hole dashboard's DNS settings. Uncheck your existing DNS servers in the left panel, and check the box beside 'Custom 1 (IPv4)' in the right panel. In the box below, type '127.0.0.1#5335', then scroll to the bottom of the page and click Save. Make sure you use a hash (#), rather than the more usual colon (:) between the IP address and port number. When the configuration page refreshes, open a browser window on any computer filtered by Pi-hole and visit a website to check that everything is working properly.

11 Flush your DNS data

If you've made changes to your DNS settings – and particularly if you've set Pi-hole to redirect bespoke addresses to computers on your network, or to use an alternative DNS service with child-friendly filtering – but you're still seeing content on connected computers that should be blocked, flush the DNS cache on your computer. On a Windows machine, open a Command Prompt window and type `ipconfig /flushdns`. On a Mac, open a Terminal window and type `dscacheutil -flushcache` and press **ENTER**, then type `sudo killall -HUP mDNSResponder` and press **ENTER**. On a Mac, you'll need to provide your password.

12 Back up Pi-hole

Now that you've made changes to your Pi-hole configuration, you should make a backup of your settings so you can reinstate them should anything corrupt the system. Click Settings in the Dashboard's sidebar, followed by the Teleported tab. Click the Backup button and save the resulting file somewhere safe. To later reinstate it, click the 'Choose file' button in the Restore panel, select the downloaded file, and click Restore. Make a backup every month or so to ensure your most recent settings, blacklists, and whitelists are reflected in the archive.

Top Tip

Minimal install

The Dashboard is available across your network, which means you can run Pi-hole on Raspberry Pi OS Lite.

Get to grips with administering Pi-hole

Explore the Pi-hole dashboard, learn how to beef up its web filtering features, and use its integrated DHCP server

MAKER

Nik Rawlinson

Esperanto-speaking, pencil-wielding, single-board computing fan who likes hyphens and remembers what that icon on the save button depicts.

nikrawlinson.com

In the last two workshops, we showed you how to set up Pi-hole to filter ads and other web content on your home network, and how to change the DNS settings so you have more granular control over the way it works. In this final instalment, we'll be taking a look at the broader Pi-hole dashboard, and showing you how the changes you make there will allow you to control what is – and isn't – available to other computers on your network. We'll integrate third-party domain lists to block distractions with less effort, and use Pi-hole's integrated DHCP server to simplify setup on connected clients.

01 Log in to Pi-hole

Although Pi-hole runs on Raspberry Pi, you can access the dashboard by pointing a browser towards **pi.hole/admin/** on any computer that uses Pi-hole to filter its web content. Other devices on your network, not filtered by Pi-hole, can also access the dashboard by adding '/admin/' to the end of the numeric IP address of the Pi-hole device. Click Login in the sidebar and provide your password. Once you've logged in, you'll notice that the dashboard suddenly becomes a lot more informative. It's worth spending some time getting familiar with its contents.

02 Understand your traffic

The uppermost two graphs depict DNS queries passing through Pi-hole. Don't be surprised if there are overnight queries, when backup services kick in and some computers may be set to update. You can see which of your devices is making the most calls at the bottom of the page, where their activity is split out into total requests and blocked requests. The 'Queries answered by' panel shows you a count of where the answer for each query was sourced from.

03 Check what's being blocked

Unless you disabled logging when you installed Pi-hole, it will make a note of every request it processes, whether permitted or blocked. Browsing the log is a good way to check whether anything on your network is attempting to make persistent connections to an unauthorised remote server, which could suggest it's hosting malware. Click 'Long-term data' and, in the section this exposes, click Top Lists. Choose a time frame by

Using third-party domain lists simplifies the task of adding comprehensive sets of addresses for social networking and other services

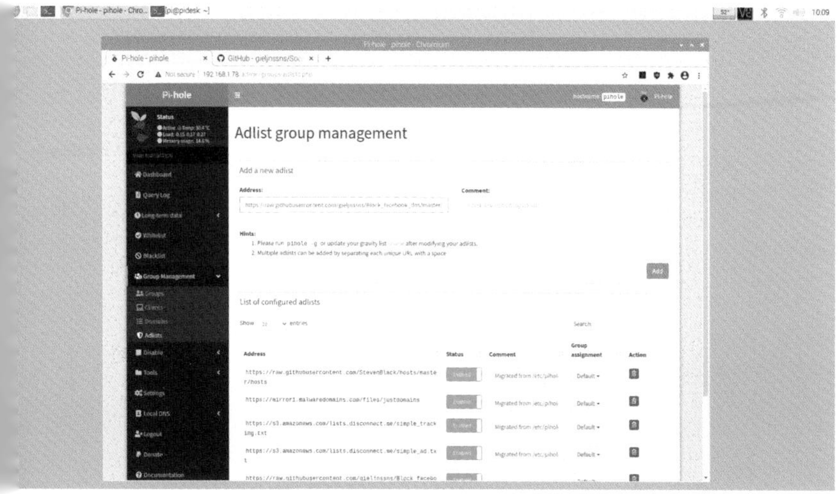

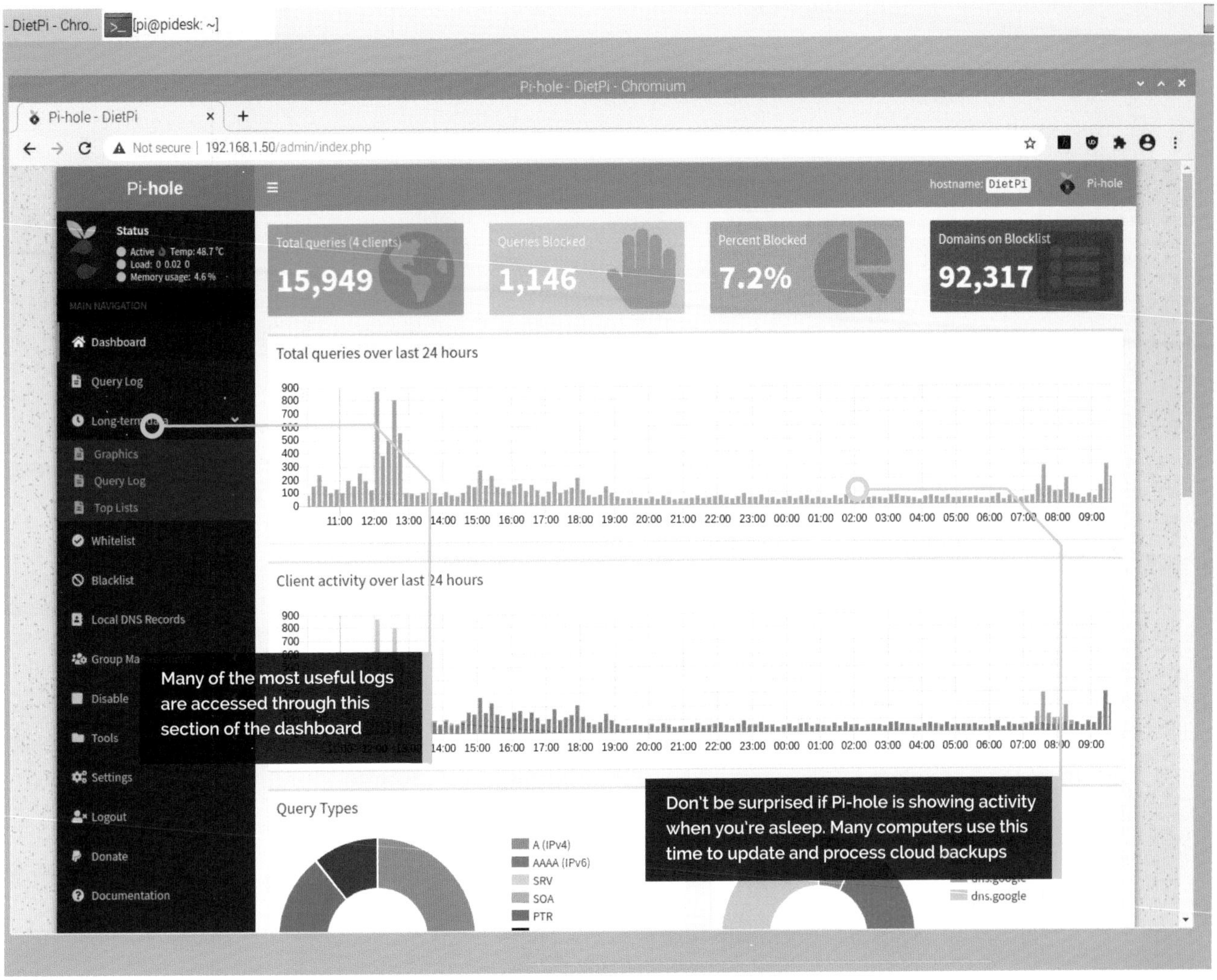

clicking 'Click to select date and time range'. Start with 'Last 7 days' and browse the entries in the Top Blocked Domains box. If anything looks suspicious, check it using a service like Kaspersky's Threat Intelligence Portal (click Lookup at **opentip.kaspersky.com**).

04 Investigate frequent callers

Should you believe the domain in question is a threat, investigate which computer – or computers – on your network is making the requests to that address. Click Query Log, immediately above Top Lists, and isolate the same time frame by once again clicking 'Click to select date and time range'. Now type the domain in question into the Search box and check the IP addresses in the Client column to see which computers on your network are insistently reaching out to it. Scan the affected computers for threats using your regular anti-malware tools.

05 Blacklist suspicious domains

If you're sure the domain is a threat, you can also block access to it through Pi-hole, which will stop any malware running on computers filtered by your Pi-hole setup from reaching it, potentially preventing them from stealing your personal data. Simply click Blacklist in the Action column against the domain you want to bar. Doing so will only affect the specific domain and subdomain as specified in the Domain column, but leave other subdomains active, so this isn't the best way to block, say, social networking sites, which frequently use different domains for browser and app access.

06 Block complete domains

So, if you want to use Pi-hole to block access to complete services, click Blacklist in the sidebar and type the last part of the domain name into the Domain box. For example, **bbc.co.uk**, but

You'll Need

- Raspberry Pi
- Pi-hole **pi-hole.net**
- Web browser

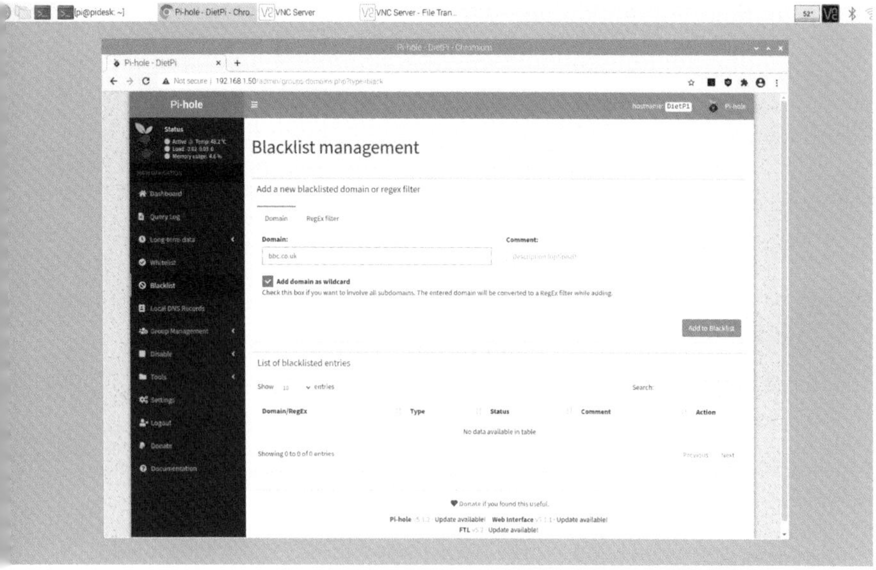

Adding a domain to the blacklist or whitelist and ticking the wildcard box will apply the same rule to all of its subdomains

Use the timeout feature to temporarily disable Pi-hole. This is useful when diagnosing connectivity issues on client devices, or for one-off visits to blocked sites

Check the status box at the bottom of the dashboard for updates to the Pi-hole software, and patch it as required

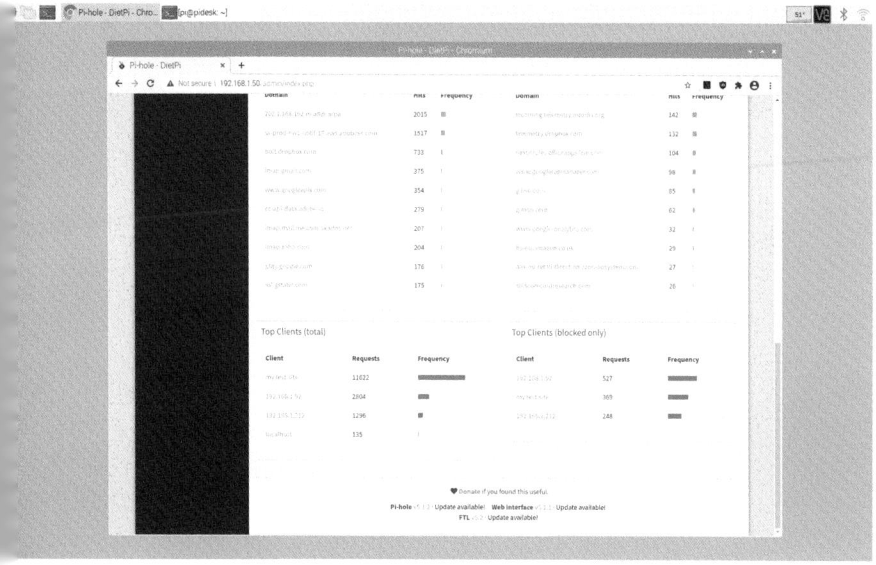

not **www.bbc.co.uk** if you want to block the BBC. Now, to block all subdomains within bbc.co.uk, including those hosting the website, feeding data to apps, serving video and so on, click 'Add domain as wildcard', then click 'Add to Blacklist'. In much the same way, if any services you need to use are being blocked by the third-party filter lists Pi-hole uses, you can specifically allow them by clicking Whitelist in the sidebar and adding them there.

07 Bolster your social blocks

Note that some services use multiple domains, which will all need blacklisting if you want to lock them out entirely. For example, Facebook also hosts data at facebook.net, fbcdn.com, fbcdn.net, fbsbx.com, and more. Rather than researching every possible domain for each social network, you can incorporate third-party lists, like those maintained on GitHub at **magpi.cc/socialblocklists**. In Pi-hole, click Group Management in the sidebar, followed by Adlists. Now paste each pihole-prefixed list into the Address box in turn (for example, **https://raw.githubusercontent.com/gieljnssns/Block_facebook_dns/master/pihole-facebook.txt**), clicking Add after each one.

08 Take a timeout

If you briefly need to visit a site that's blocked by Pi-hole, or you're trying to diagnose a problem that you suspect the filter might be causing, you can temporarily disengage the whole system, rather than trying to work out which domains you need to unlock. Click Disable in the sidebar, then choose how long you'd like to switch off for. The default options are 10 seconds, 30 seconds, and five minutes, but you can also pause filtering indefinitely, or click 'Custom time' and enter your preferred duration. This exposes every device previously filtered using Pi-hole, not only the one through which you're accessing the dashboard.

09 Assign network addresses

So far, we've assumed you're going to continue using your router's DHCP server to assign numeric addresses to clients on your network, which we'd recommend if you want to use Pi-hole on just a few of your devices. However, if you intend to use it to filter content for every machine you own, it makes sense to also enable its DHCP server. Doing so means they'll automatically use Pi-hole as their DNS server, too, which simplifies configuration and means any new computers – and devices like voice assistants and smart TVs – will be secured as soon as they connect to your network.

10 DHCP server

Click Settings, then click the DHCP tab. Click the checkbox beside 'DHCP server enabled', then use the From and To boxes to specify the range of addresses that Pi-hole can hand out. Unless you have a lot of devices on your network, you should be safe to leave the defaults as they stand, giving you capacity for 50 devices. Check that the address of your router, as specified in the Router step below, is correct. If you don't know what your router's address is, open a Terminal window and type:

```
ip route | grep default
```

The router address appears immediately after 'via'.

11 Disable router-based DHCP

You shouldn't have two devices handing out IP addresses on your network, so log in to your router's admin pages through the browser

> You shouldn't have two devices handing out IP addresses on your network

by typing in the address you obtained using the **ip route** command in the previous step. Navigate through its various screens until you find the DHCP setting. If this is protected, the administrator password is often specified on the back of the router itself. Clients on your network will seek out a replacement DHCP server automatically and, with your router no longer providing this function, should find your Pi-hole installation, which will assign them the address they require.

12 Keep Pi-hole updated

The status bar at the bottom of every dashboard page will warn you if your installation is out of date. At the moment, it's not possible to update the system via the graphical interface. So, if you spot a red, pulsing 'Update available!' beside the Pi-hole, web interface or FTL version numbers, open a Terminal window on the Raspberry Pi running Pi-hole and enter:

```
pihole -up
```

Pi-hole will check for the latest available version, then download and install it automatically.

Top Tip

Terminal tweaks

TPi-hole can be tweaked and updated at the Terminal, too. Check **magpi.cc/piholecommand** for a comprehensive list of commands.

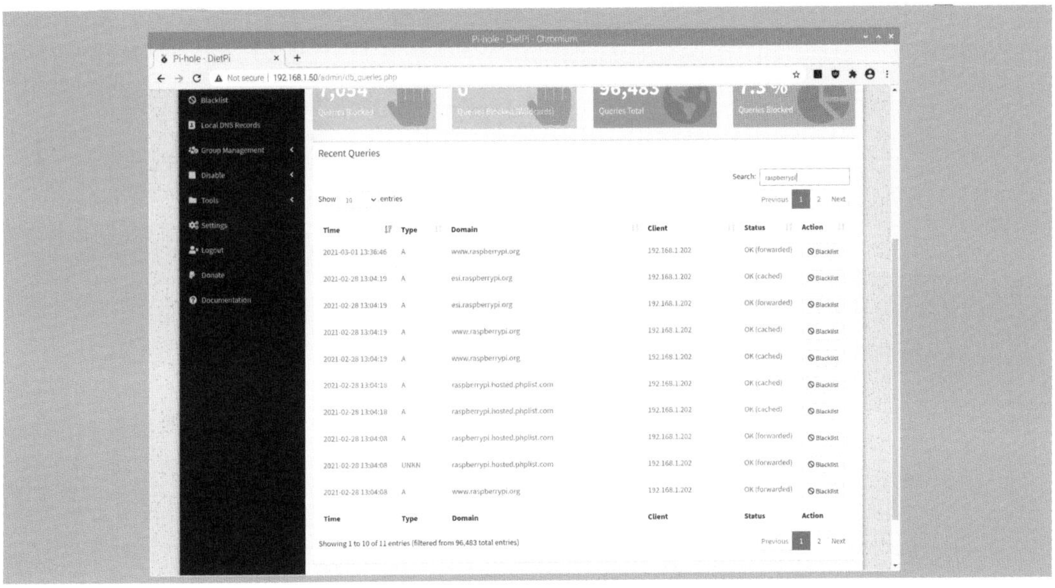

When you've queried your logs, you can add domains to your blacklist directly by clicking the appropriate button

Top Tip

Double servings

Run two Pi-hole servers on your network and if one falls over, devices can switch to the other until it returns.

Build an arcade machine: Get the parts

If you've ever wanted to build your own arcade machine, here's your guide. This month: the parts you'll need, how to choose them, and where to buy them

MAKER

K.G. Orphanides

K.G. is a writer, maker of odd games, and software preservation enthusiast. Their family fully supports the idea of an arcade machine in the living room.

@KGOrphanides

Over the coming months, we'll go through the process of sourcing, building, connecting, and installing a Raspberry Pi-based arcade cabinet.

While you can restore and convert a former JAMMA cabinet for use with Raspberry Pi, or build a cab entirely from scratch, we'll be taking the flat-pack route. This lets you build the cabinet of your dreams relatively easily, somewhat cheaply, and without recourse to full-on home woodworking.

This tutorial series will use an LCD screen due to the inconvenience of sourcing and potential issues with installing a CRT model, which carries the risk of a dangerous electric shock if not correctly discharged.

01 Choose your cabinet style

If you're after a classic upright one- or two-player cabinet, then you'll want either an all-in-one model or a 'bartop' cabinet with a pedestal or stand. Bartop cabinets can also be bought without the optional stand and placed on a table.

Flat 'cocktail' or 'coffee table' style cabinets are available in models for between one and four seated players and often use a vertically oriented screen, which can be split by software into two horizontal views for multiplayer games.

Other models include seated upright cabinets (often designed to take very large screens), angular tabletop models, and mini-bartops with 10-inch displays for those short on space.

02 Big screen glamour?

The size of your screen dictates the size of your cabinet, and vice versa. Before you start shopping, work out where you want the cabinet to live, and take height, width, and depth measurements.

If you're working with a 19-inch monitor, you'll likely get a bartop cab that's a little under 50 cm wide. This is the most practical choice if available space is limited. A 22-inch screen translates to a cabinet of a little under 60 cm, and a 24- or 25-inch screen means a cabinet width of a bit under 65 cm. You're generally fine fitting a smaller screen in a larger cabinet, but the end result won't look quite so polished.

Check the internal measurements of the cabinet against those of the monitor, including its bezel.

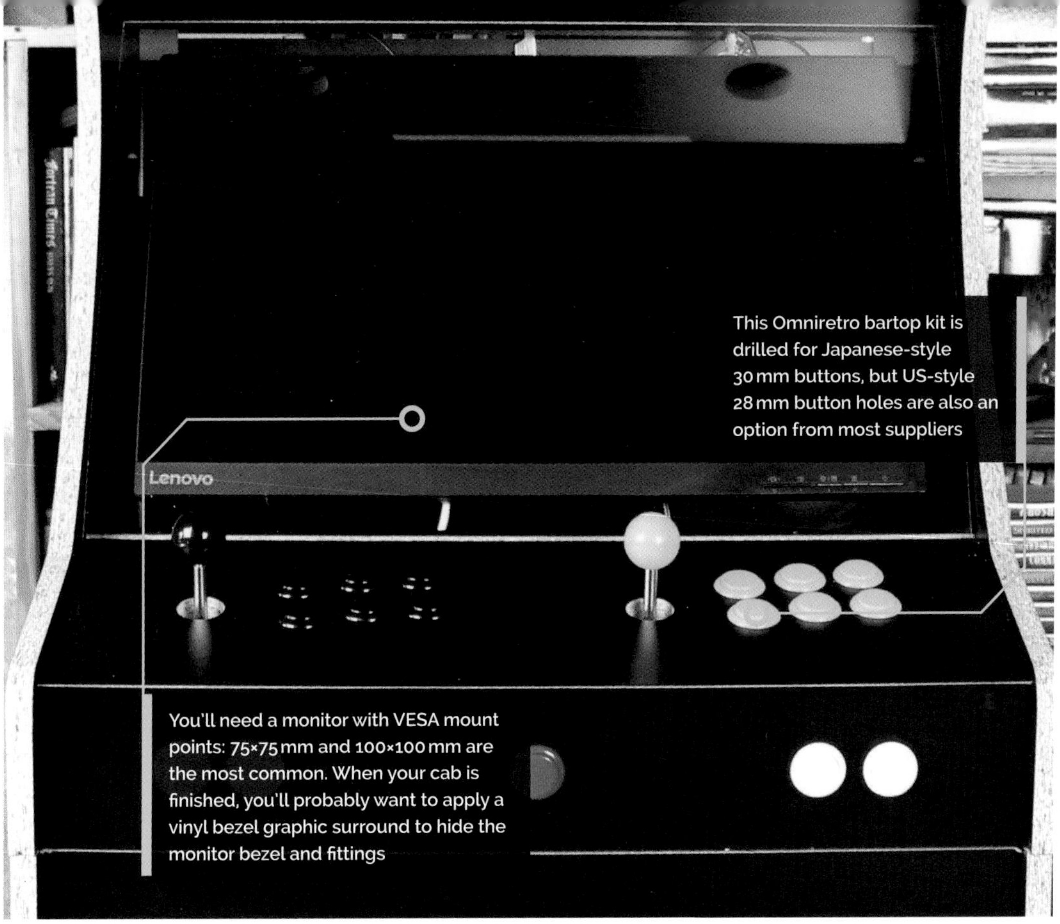

This Omniretro bartop kit is drilled for Japanese-style 30 mm buttons, but US-style 28 mm button holes are also an option from most suppliers

You'll need a monitor with VESA mount points: 75×75 mm and 100×100 mm are the most common. When your cab is finished, you'll probably want to apply a vinyl bezel graphic surround to hide the monitor bezel and fittings

Top Tip

Button positioning

We're going with a six-button Japanese-style layout. Check out **magpi.cc/joysticklayout** to see some alternatives.

03 A good fit

Depending on the era of games you want to play, a large 1920×1080 widescreen display may not be the most authentic choice, but it is the most flexible, and modern emulators handle HD displays well.

Most cabinets have a VESA mount, usually in the form of a monitor support bar drilled for 75×75 and 100×100 mount points. Make sure your monitor has mounting points that match.

Finally, ensure that your monitor will work with Raspberry Pi: anything with a standard HDMI input should be fine, but older DVI and VGA displays require inconvenient adapter arrangements.

04 Materials

Self-assembly cabinets are usually made in MDF, but laminate, melamine, and veneer finishes are also widely available.

MDF swells badly if exposed to water, so if you're going to have drinks anywhere near your cabinet, a water-resistant finish is strongly recommended. If you buy an untreated MDF kit, apply and sand down between multiple coats of an MDF-specific solvent-based primer, then paint it to your heart's content, ideally with oil-based paint.

18 mm MDF is common, but you'll find cabinets in anything down to 10 mm for budget models. 18 mm or thicker construction materials may require a longer shaft or extender for your joystick. If in doubt, talk to the kit's supplier.

05 Finish and decoration

Regardless of the materials used, you'll probably want some plastic edging strip. This plastic trim helps to protect the edges of your cabinet, makes it easier to clean, and looks a lot more professional than exposed MDF edges.

Two types are popular. T-Molding is more secure but requires a slot to be cut for it to clip into – some DIY kits have ready-cut slots for this purpose, but budget models frequently do not.

U-Molding just clips over the edge. Cabinet makers will usually tell you how much moulding

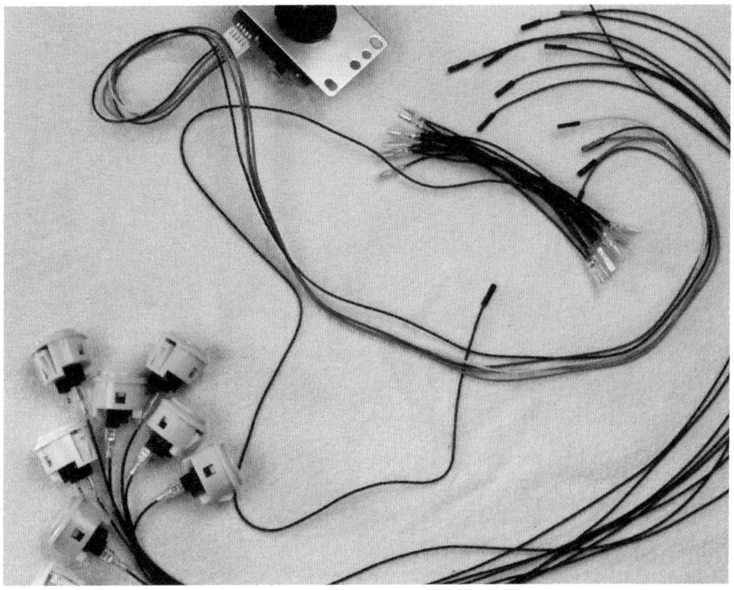

▼ You can get kits containing all the joysticks, buttons, and connectors you'll need; just make sure your button and cabinet hole sizes match

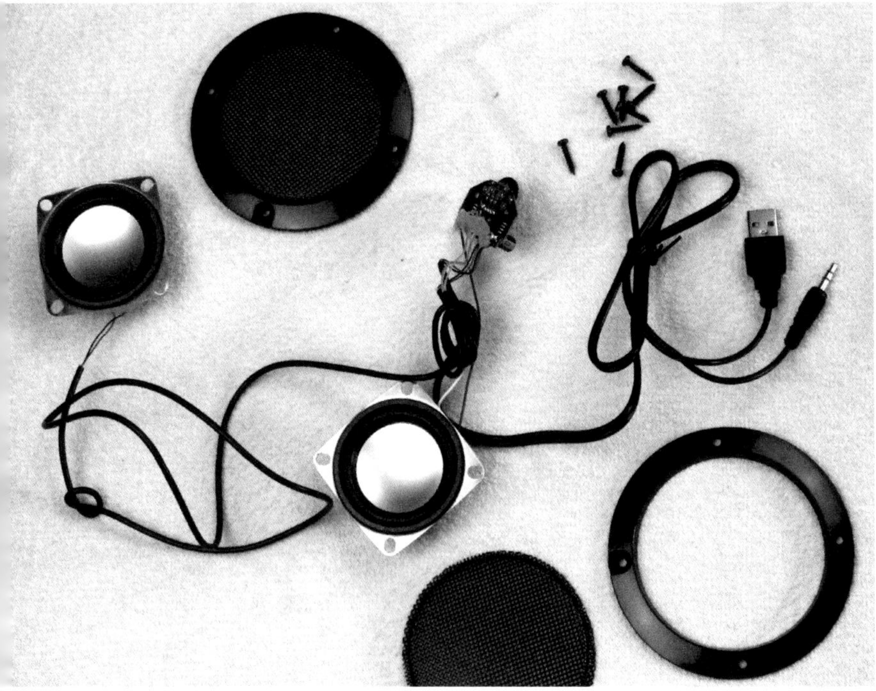

▲ A variety of compact bus- and mains-powered amp and speaker kits are available: this one takes power from the USB port and audio from the 3.5mm port

their kit will need and can usually supply the required quantity and type of edging.

Many arcade cabinet suppliers also sell a range of decorative and protective graphical vinyl sticker wraps. These should be applied with care to an appropriately finished surface (check with the sticker manufacturer for any finish requirements).

06 A giant screen protector

To protect your screen and create a flush finish, you can – and should – opt for an acrylic (polymethyl methacrylate, also known as plexiglass) screen protector. Again, this is something most self-assembly kits are designed to take and the majority of retailers will happily sell you one as either a standard part of the kit or an optional extra. Make sure you do opt in, as cutting your own plexiglass to precise dimensions can be a pain. Toughened glass and UV-resistant polycarbonate can also be used. You may need to add some standoffs to stop front monitor buttons being pressed by the screen protector.

07 The marquee club

Also included in kits as a matter of routine is a strip of acrylic for your cabinet's top marquee. You'll probably want to get a backlight-ready vinyl marquee (available from print shops, arcade suppliers, and on Etsy) to stick to this, but you could also decorate your own.

Sample shopping list

Here is an illustrative price list. The prices include VAT but not shipping or additional costs.

Item	Price
24-inch LCD monitor	£125.00
Bartop cabinet	£170.00
Bartop stand	£100.00
10 m T-Molding	£25.00
Acrylic control panel guard	£25.00
Two-player USB joystick + button kit	£70.00
Amp, speaker & cover kit	£25.00
Amp power supply	£12.00
Printed marquee	£6.00
LED strip lighting	£15.00
Molex power adapter for LEDs	£15.00
5-way plug bar	£15.00
TOTAL	**£603.00**

While you're at it, you may wish to get acrylic or metal panels to surround your buttons and joystick. These can be decorated, and protect your cabinet's surface, as well as providing a smoother feel. Button layouts tend to be standard, but these should ideally be bought from the same supplier as your kit for the best fit.

08 Raid the button tin

We'll be building a cabinet with an eight-way joystick and six 30 mm buttons, plus Start and Select buttons, for each player. A variety of alternative sizes and brands are available, with Sanwa perhaps being the most recognisable. You can order a cabinet with holes for extra side buttons if you're into digital pinball.

An easy cross-platform connection solution is a USB arcade encoder. Models by Zero Delay and Xin-Mo are popular, but the I-PAC 2 keyboard encoder has slightly lower latency.

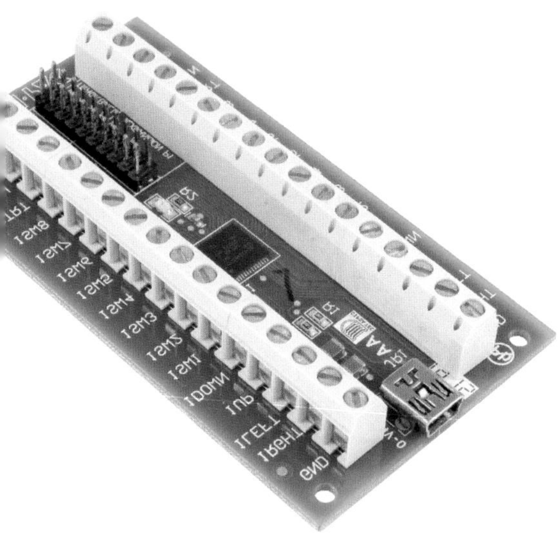

If you want to use USB, the Ultimarc I-PAC 2 encoder is a popular choice that'll work with most computers. Check out **magpi.cc/ultimarcgit** for advanced configuration

09 Pick a driver

You can connect controls to Raspberry Pi's GPIO, using either the Adafruit Retrogame (**magpi.cc/adaretrogame**) or mk_arcade_joystick_rpi (**magpi.cc/mkjoystick**) drivers – we'll be using the latter.

Arcade joysticks generally use a five-pin JST connector, while non-illuminated buttons each have a pair of quick-connect spade connector fittings, one of which must go to ground. Spade to DuPont GPIO cables are uncommon, but can be bought either individually or as part of a kit from specialist retailers such as SmallCab. Illuminated button kits are available with an extra external PSU.

> LED strip lighting is a popular choice for marquee panels

10 The sound of success

It's a good idea to order your cabinet with a couple of pre-drilled speaker holes and covers to go over them. The most common option for audio is an externally powered stereo amp, connected to Raspberry Pi's 3.5mm port, and 10cm/4-inch speakers, but USB-powered kits are also available. If you have one lying around, you could also consider mounting a compact USB sound bar behind your speaker grilles.

11 More power, Igor!

A major advantage of this kind of arcade machine build is that there are no internal power supplies to bother with. There's enough space to mount a plug bar inside most cabinets, and you can use this to power the monitor, Raspberry Pi, and any extra transformers required for lights or speakers.

Where to buy

There are a number of UK and EU retailers specialising in self-assembly arcade cabinets and components. While it's easiest to get everything in one place, you have to mix and match for specialist components such as GPIO-compatible wiring looms.

- **Arcade World UK** – arcadeworlduk.com – supplies a wide range of kits and components; discount codes available for most non-furniture items
- **Bitcade** – magpi.cc/bitcadekits – UK arcade machine maker that also supplies kits
- **Omnireto** – omniretro.com – Spanish firm with a notable budget range
- **Rockstar Print** – rockstarprint.co.uk – custom marquee and wrap printer
- **SmallCab** – smallcab.net – French supplier of arcade kits and hardware including GPIO-friendly wiring

Warning!
Paint and dust

When sanding, sawing, or painting, be sure to use appropriate eye and breathing protection in a well-ventilated space.

magpi.cc/diysafety

LED strip lighting is a popular choice for marquee panels, but you'll need to buy a Molex power adapter to go with it, or repurpose a PC power supply. You can run a plug lead out of the back or optionally install an external power socket and switch, if you're comfortable with simple electrical wiring.

12 Room to build

Before you start ordering, consider not only the space you have to house your cabinet, but also how much room you have to build in. Don't get an untreated MDF cabinet unless you have a large, ventilated (and paint-resistant!) space where you can apply primer to each part, as well as appropriate eye and breathing protection.

You'll want to source durable joysticks and buttons for your arcade machine

Build an arcade machine: Assemble your cabinet

Once your arcade cabinet kit arrives, it's time to put everything together

MAKER
K.G. Orphanides

K.G. is a writer, maker of odd games, and software preservation enthusiast. Their family fully supports the idea of an arcade machine in the living room.

@KGOrphanides

In this tutorial, we will assemble an arcade cabinet, fit controls, and mount a monitor. You should follow the video or written assembly instructions for the model you buy, but we'll go through the process so you know what to expect and how to handle the awkward bits.

Kits don't necessarily come with the screws and bolts you'll need to attach parts such as speakers, speaker grilles, and monitors, so check that you have all the hardware you'll need before you start.

Our cabinet is an Omniretro Bartop Arcade King with a stand (**magpi.cc/kingbartop**), made of 16 mm black melamine laminate, and we are using a 24-inch monitor.

01 Lay out your parts

MDF and melamine laminate are light, cheap, and sturdy when assembled, but can be susceptible to damage if dropped or pivoted hard on an edge or corner.

Make some space and put down towels to protect the cabinet parts and your floor from one another. If your unit consists of a separate bartop and stand, build them one at a time. Read or watch the manufacturer's instructions and make sure that you have all parts, fixings, and tools to hand before you start.

02 Preparation

Assembly varies from brand to brand. If access to the assembled cabinet is restrictive, you may have to fit your buttons and joystick to it before you put it together.

Similarly, attach speakers to the inside of the marquee bottom and speaker grilles to the outside before you assemble the cabinet. If you're working with laminate, mark up the screw positions through their holes with a paint pen and use a 3 mm bit to drill pilot holes.

If you've already decided on your marquee, control panel and bezel graphics, your life will be easier if you apply these to their acrylic sheets before assembly (we'll be looking at this in detail in a later tutorial).

> " Put down towels to protect the cabinet parts and your floor from one another "

03 Assembly

If you're comfortable with self-assembly furniture, an arcade cabinet shouldn't present too much trouble, but a second person can be helpful for fitting and moving awkward parts.

Ours has a control panel with a hinged access door beneath it, so we attached this hinge first

It's a good idea to fit your speakers and grilles before assembling the cabinet, but it's possible, if fiddly, to do it afterwards

We have put U-moulding onto the edges of the cabinet now to protect them, loosely secured with standard double-sided tape at the ends. We'll re-secure this properly after decorating the cab

Top Tip

Snap-out

Snap-in buttons can be hard to remove without damage. The ButterCade Snap-out tool for push-buttons (**magpi.cc/snapout**) is a plastic device to help with this.

You'll Need

- Screwdrivers, spanners, Allen keys, crimping tool
- Cordless drill
- Drill bit set. Screwdriver bits, drill bits, countersinks, tank cutters
- Additional bolts, screws, female spade connectors (to mount components)
- Dremel (recommended) and 3 mm drill
- Paint pen (silver if you have black laminate, black for MDF)
- Old towels or sheets to protect parts
- Foam cleanser and microfibre cloths (to clean your cabinet and acrylics)

The underside of a Sanwa JLF-TP-8YT joystick. Note the e-clip securing the central shaft

then set the panel aside. We then attached the hinge for the bartop's rear access door and base, lined this part up with the cabinet's hood-like top, and bolted all of these parts to one side panel laid on top of them.

Lining bolts up with pre-drilled holes for this kind of build can be fiddly. If you have trouble, screw the bolts through the side panel until they're protruding, and use them to help find the correct positions.

04 The control panel

With one side now in place, slide in the control panel and bolt it to the same side as the other parts. Next, attach the marquee bottom that houses the speakers, which should already be mounted at this point.

With this model, we then close the latch on the rear access door and carefully flip the entire cabinet over onto the now-secure side panel. This is the best time to slide the marquee and screen acrylic panels into place. If you've not already applied graphics to them, leave their protective film on – it's easy to peel off later.

We now position the second side panel. We recommend again screwing in the bolts until they just protrude from the opposite side to help you lower the panel securely and accurately onto its pre-drilled holes.

05 Feel the power

Drill a hole at the back of your bartop and run the power bar's cable out through it to connect directly to a plug socket.

Some suppliers will wire a socket and bar for you, but note that international plug standards differ. Use a plug bar that can be surface-mounted inside of the cabinet.

While you're back there, cut a hole to accommodate a booted Ethernet cable or, more tidily, a screw-down Ethernet extension port. This will make Steam Link game streaming easier.

06 Extending your shaft

If your cabinet is over 16 mm thick, you'll want a longer than standard joystick shaft. Shafts are easy to swap, but watch out for parts dropping out.

Like most sticks, our Sanwa JLF-TP-8YT's shaft is held in place at the bottom by an e-clip. Hold the unit upside down, press on the bottom of the shaft with your thumb, and use a small flat-head screwdriver in your other hand to pull the clip off, using the slots in it. Pull the old shaft gently out from the top and push the new one in, carefully setting the pivot at the top and the spring and black plastic actuator at the bottom into place.

Use a thumbnail to depress the actuator and slide the e-clip back into place. You can also use pliers or your screwdriver to help push it on. For a demonstration, see this YouTube video on changing joystick shafts: **magpi.cc/joystickshafts**.

To fit the VESA mount, place the cabinet face-down, then put the mounted monitor face-down on the front acrylic screen. Use a tape measure to help with positioning

Warning!
Mains electricity & power tools

Be careful when handling projects with mains electricity. Insulate your cables and disconnect power before touching them. Also, be careful when using power tools during this build.

magpi.cc/drillsafety
magpi.cc/electricalsafety

Cable tidy

Cable lacing is a cable management technique where a nylon cord is used to bind wires together. It can be used to create incredibly neat builds, like this Arcade Stick by Gordon Hollingworth, Raspberry Pi's Chief Product Officer.

Gordon learnt to cable-tidy this way as an apprentice for the MOD. "Tying the knot has to be done in a very specific way to avoid it looking untidy," he tells us, "basically a capital offence in the apprenticeship!" Gordon's cables have knots regularly at 1cm, which keeps them smart. "We learnt this way because when you put a box into a plane or tank with some equipment in it, the vibration will shake apart pretty much any connection in the first hour. So this was the way it was done when electronics was more about wires connecting things than PCBs."

You can buy nylon cord and learn more from RS Components (**magpi.cc/cablelacing**).

Snap-in buttons are held in place by plastic clips. Connect your DuPont GPIO cables first to make internal wiring easier

> There's room to slide the screw slots on most joystick mounting plates

07 Installing your joystick

Two plastic dust washers come with Sanwa joysticks. Slide one onto the shaft before you mount the stick onto the underside of your control panel.

When mounting your joystick, position it, mark up the position of the top right screw-hole on the joystick's baseplate with a paint pen, and drill a pilot hole, being careful not to go all the way through.

Attach your joystick by that screw, make sure it's centred, and mark up the next hole or holes. There's room to slide the screw slots on most joystick mounting plates, so you've got a bit of wiggle room when it comes to the final fit.

Don't worry too much about the orientation of your joystick – position it where it won't get in the way of the rest of your wiring. These are nominally designated up, down, left, and right positions; you can reassign these through wiring and in software.

Finally, slide the second dust washer onto the shaft on the other side and screw the joystick's ball on.

08 World of buttons

Snap-in buttons are ideal for thick wooden cabinets – plastic clips hold them in position inside the holes drilled for them. If you have an acrylic cover for your control panel, the buttons will hold it in place.

It's a good idea to attach your spade connectors to DuPont GPIO jumper cables before installing them, but you'll have to connect the shared ground cable after they're in place. We wired GPIO to the right and shared ground to the left connector on each button, but it doesn't matter which goes where.

Where you have longer stretches between buttons, skip a connector on the ground chain to give yourself some extra cable to play with.

You can label the end of each GPIO cable for later ease of connection to Raspberry Pi, but they're not too hard to trace in most cabinets.

Top Tip

Foot the bill

To help the cabinet stand on an uneven floor, you can fit four rubber feet to its underside.

▲ To make it easier to line up the sides of your cabinet with their pre-drilled recieving holes, partially screw in each bolt until a couple of millimetres protrude on the far side

"If you find that you now can't reach or fit a part, don't panic"

▼ Before construction, lay out the parts of your bartop on some old towels to protect them

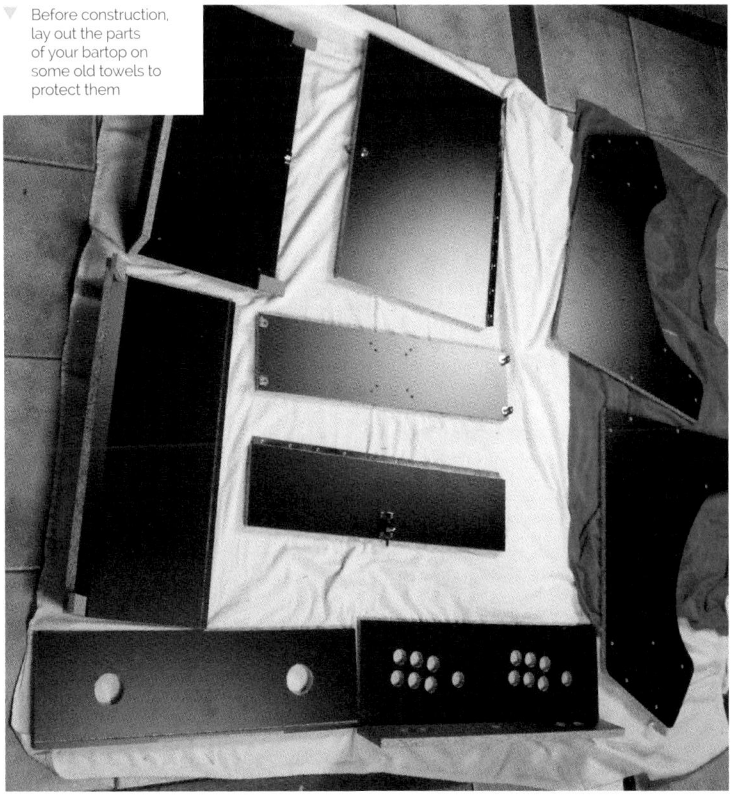

09 Bolt screen to VESA mount

Screen mounting can be fiddly. Most cabinets come with a baton-like wooden VESA mount that's designed to be screwed into place from the inside. Start by bolting your monitor to the mount. Unless you're working with a specialist cab designed for giant screens, you'll be using a 75 mm or 100 mm VESA mount. These usually take M4 bolts and have a depth of 10 mm. So if bolts aren't included, you'll need four, at a depth of 10 mm plus the depth of your mount, although you can get away with shorter if you countersink them.

10 Screw VESA mount to the cabinet

Place the bartop face-down on the ground, protecting it with a towel. Take the protective plastic off the acrylic on the inside of the cabinet. Clean the acrylic with a microfibre cloth and anti-static foam cleanser

Lay the monitor, attached to the cabinet's VESA mount, face-down on the acrylic inside your cabinet. On the interior sides of the cabinet, mark up the position of the holes in the brackets on each edge of the VESA mount. Remove the mount, drill pilot holes, then replace and screw down the display and its mount.

If your monitor has a front power button, you can use adhesive chair leg floor protectors as soft spacers to stop it from being pressed by the acrylic screen.

11 Don't panic

If you miss a stage in your build and find that you now can't reach or fit a part, don't panic. Speakers – and any other components in need of securing – can be attached internally using strong double-sided foam tape.

Most external parts can be drilled and fitted in situ. If you want to deal with decoration last, then you can sometimes pop out your acrylic panels or, better, remove one side and reattach it.

As you'll see from the photos, we have temporarily applied U-moulding to protect the edges of the cabinet. U-moulding is easy to remove and refit or replace, assuming you don't glue it down, but T-moulding is a little harder to remove cleanly.

We're now ready to connect Raspberry Pi. That will be covered in the next tutorial.

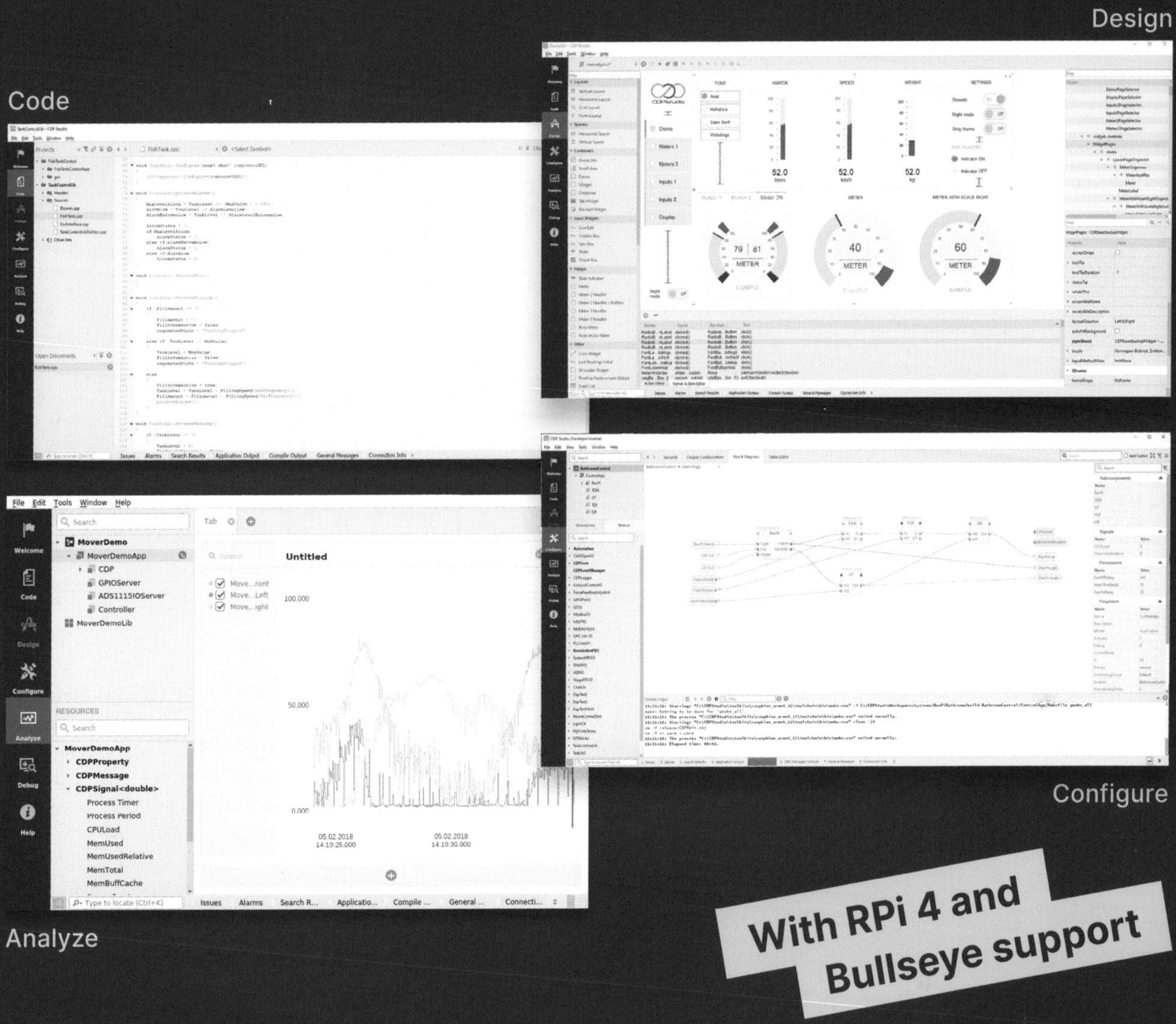
Design
Code
Configure
Analyze
With RPi 4 and
Bullseye support

PROFESSIONAL CONTROL SYSTEM DEVELOPMENT TOOL

Home projects made easy.

CDP Studio, a great software development tool for your home projects. Build systems for Raspberry Pi, use C++ or NoCode programming, open source libraries, out of the box support for GPIO, I2C, MQTT, OPC UA and more. Create beautiful user interfaces. Built for industrial control system development, **FREE for home projects**.

cdpstudio.com

Tel: +47 990 80 900 • info@cdptech.com

CDP Technologies AS // Hundsværgata 8, 6008 Ålesund, Norway

CDPStudio

Build an arcade machine: Command and control

We've assembled our cabinet. Now it's time to put Raspberry Pi to work with the Recalbox arcade OS

MAKER

K.G. Orphanides

K.G. is a writer, maker of odd games, and software preservation enthusiast. Their family fully supports the idea of an arcade machine in the living room.

@KGOrphanides

With our arcade cabinet built, it's finally time to get emulating with Raspberry Pi. We're using the Recalbox emulation distro for this project, which has excellent GPIO arcade controller support, a slick front end, and a handy web interface to help you configure and manage it.

The RetroPie distro is a popular choice for arcade machines, and adds Steam Link support, but requires manual installation and pull-up switch reconfiguration to get GPIO arcade controls working.

01 Wire up your controls

Last month, when we added buttons to our cabinet, we recommended attaching the spade-to-DuPont cables that will connect to Raspberry Pi's GPIO before inserting the buttons. If you didn't, it's time to open the back of your cab, grab a torch, and get in there to fit them.

Connect a spade-to-DuPont cable to each button and connect a shared ground cable to each of the left and right button banks. Where you have longer stretches of buttons – for example between the central hot button connected to player one's rig and the player one start button – it's a good idea to skip a connector on the ground chain to give yourself some extra cable to play with.

Plug the 5-pin cable into the joystick. Looking at our Sanwa stick from below, the bottom-most pin, which connects to the black cable strand on standard-coloured 5-pin wiring harness, is ground.

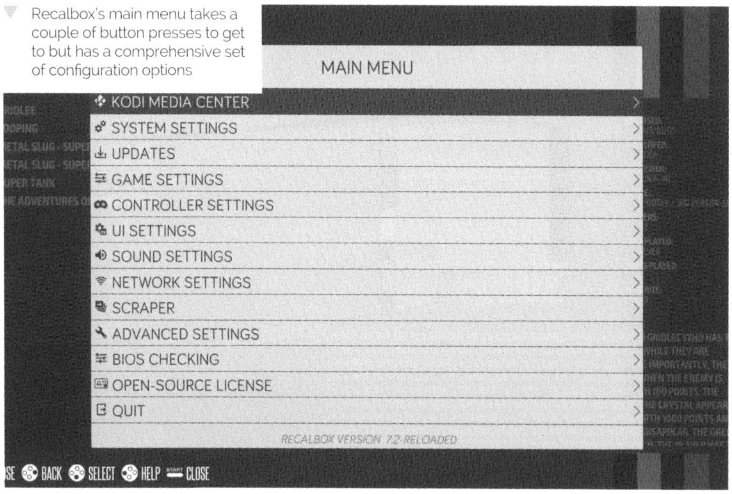

Recalbox's main menu takes a couple of button presses to get to but has a comprehensive set of configuration options

02 Connect to GPIO

This is the fiddly bit. We suggest using a case for Raspberry Pi that fully exposes the GPIO pins. The GPIO wiring diagram shows which buttons, directional controls, and ground connections should be attached to each pin. While buttons and controls can be reconfigured in software, ground cannot. Our setup uses a total of 25 GPIO inputs, plus four ground connections. Input 25 is for a dedicated hotkey button.

03 Install and power up

Open Raspberry Pi Imager, connect your microSD card writer, and Choose OS > Emulation and game OS > Recalbox and the version of Recalbox that matches your Raspberry

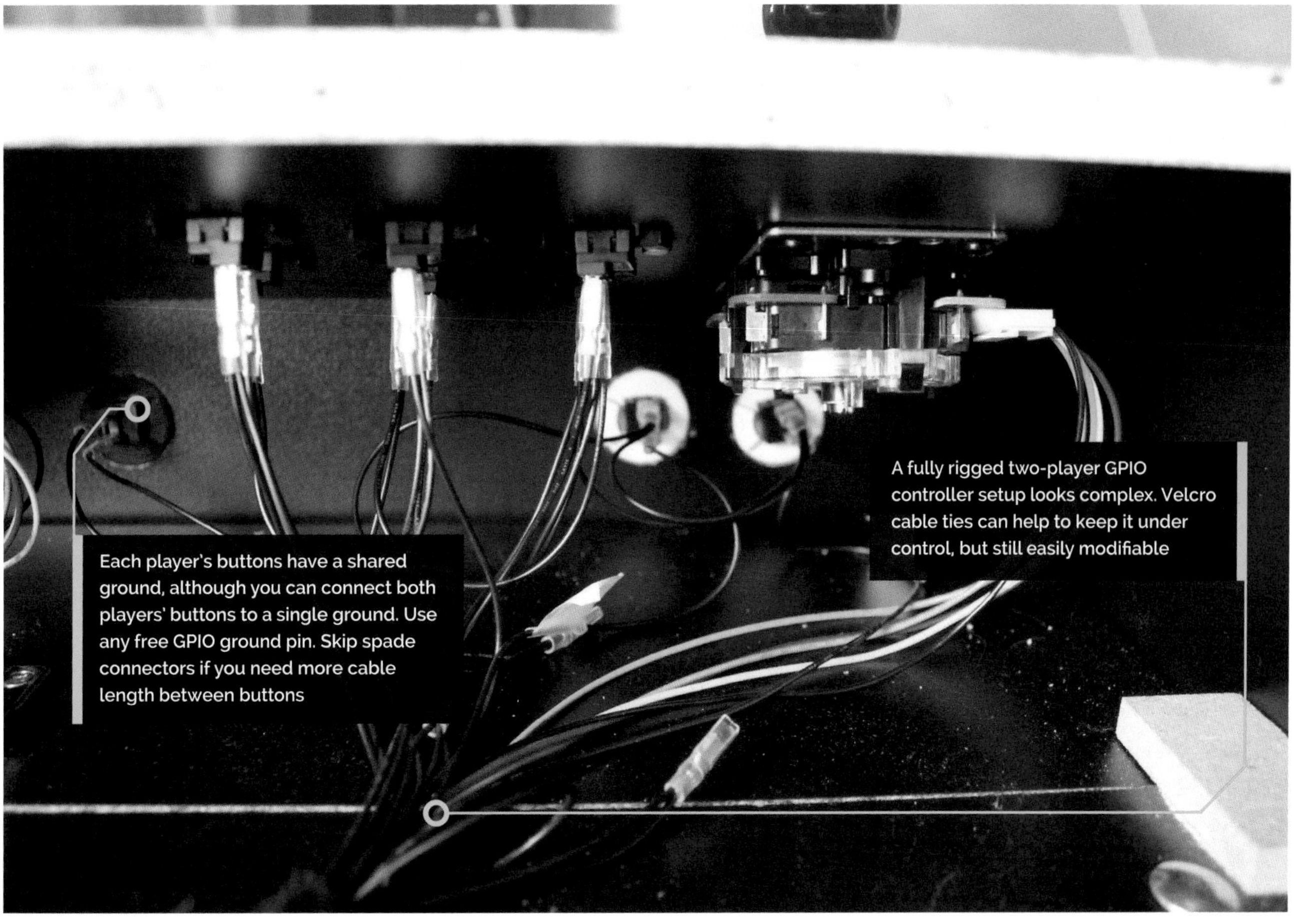

Each player's buttons have a shared ground, although you can connect both players' buttons to a single ground. Use any free GPIO ground pin. Skip spade connectors if you need more cable length between buttons

A fully rigged two-player GPIO controller setup looks complex. Velcro cable ties can help to keep it under control, but still easily modifiable

Pi. Click Write and wait for the image file to be written to the microSD card. When Imager has finished, remove the microSD card and insert it into the Raspberry Pi in your arcade build. Connect the cabinet's monitor and speakers to Raspberry Pi. Plug in a keyboard on a long cable. Plug in Raspberry Pi's power and it will boot to Recalbox's EmulationStation interface, which you can immediately navigate using the keyboard. However, we still have to enable our GPIO arcade controls, wireless networking, and other configuration options.

04 Connecting Recalbox

Recalbox has SSH and Samba enabled by default, as well as a web interface available via your browser on **recalbox.local**. Recalbox should appear on your network as RECALBOX (File Sharing).

> "Recalbox has SSH and Samba enabled by default, as well as a web interface"

A wired Ethernet connection will give you immediate access to these. If you don't have one, press **ENTER** to open the menu, scroll to Main Menu, and select it with **A** on the keyboard, then select Network Settings, enable WiFi, select your SSID, and then select 'Wifi Key' to enter your password. Recalbox only has a root user. The default username is **root** and the password is **recalboxroot**.

05 Configure Recalbox

You can access Recalbox's config file – **recalbox.conf** – by connecting via SSH, by browsing to the system directory in the Recalbox (File Sharing) Samba share, by pressing **F4** and then **ALT+F2** at the cabinet to exit to the console, or by going to **http://recalbox.local/** and selecting the **recalbox.conf** tab in the left-hand menu pane.

Under 'A – System options, Arcade metasystem', remove the semicolon that comments out `emulationstation.arcade=1`. This will make the arcade category the first entry in Recalbox's EmulationStation interface.

You'll Need

- Spade to DuPont cables
- Spade to DuPont shared ground cables
- Joystick to DuPont cables
- At least one Neo Geo Classics Collection game **magpi.cc/ironclad**

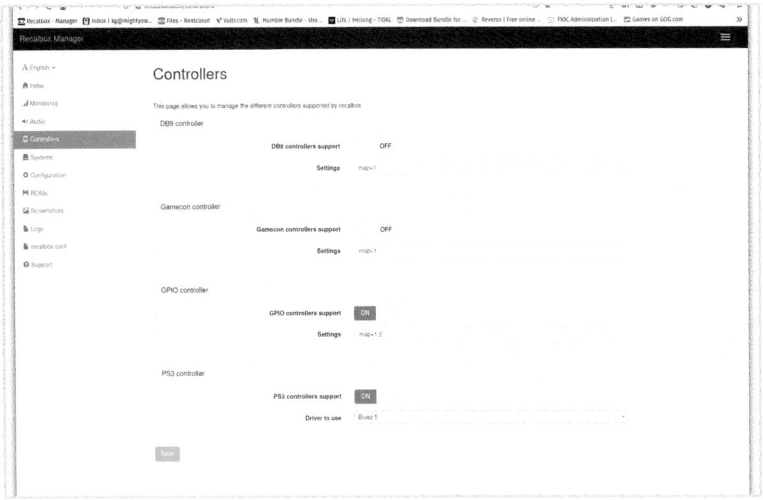

▲ A web interface at http://recalbox.local gives you control over almost every aspect of your arcade machine's setup

Under D2 - GPIO controllers, set `controllers.gpio.enabled=1`. Save your changes and, at the arcade cabinet, open the menu, go to Quit > Fast restart Recalbox.

Top Tip

USB controls

To convert your controls to USB, use a Xinmotek board (**magpi.cc/xinmotec**) instead of connecting to GPIO.

06 Optional: Take control

Recalbox will now automatically detect GPIO controllers and, if all your buttons are wired as it expects, will already have the correct button configurations. Use the bottom-left button (B) to select options and the bottom-centre button (A) to go back. Left and right navigate between systems; up and down navigate between games or options within a menu. Press Start to open the configuration menu.

If your buttons aren't connected in that order, or if you prefer an alternative layout, open the menu and go to main menu > controllers settings > configure a controller. Press down to skip an entry that you don't have buttons for. If you don't have a hotkey button for one or more players, set it to Select.

07 Sounds good

If you have no sound, open the menu, select sound settings, and check the output device. We had to switch to 'headphones – analog' output to use our cabinet's speakers, connected to the 3.5 mm output on Raspberry Pi.

Recalbox plays background music all the time by default. This is charming, but a bit much for a cabinet that lives in the sitting room. Switch the Audio Mode to 'Video sound only' – to only hear the splash screens on boot – or 'No sound'.

If you prefer, you can add your own music by copying it to Recalbox's **share/music** directory.

08 Getting to know Recalbox

Recalbox comes preloaded with a number of freeware and open-source games. Because we enabled arcade mode, this category appears first. There are already four games loaded into it.

Select the category by pressing button B and scroll through them with the joystick. Gridlee, released in 1982, looks great for the era. Press B to load it.

Press Select to add credits and press Start when you're ready to play. When you've had enough, press the hotkey button and Start together to quit back to the Arcade menu.

> "Press the hotkey button and Start together to quit back to the Arcade menu"

You can press A to go back to the top menu, and use the joystick to navigate up and down through the list. But it's easier to use the right and left directional controls to navigate through each console's full library.

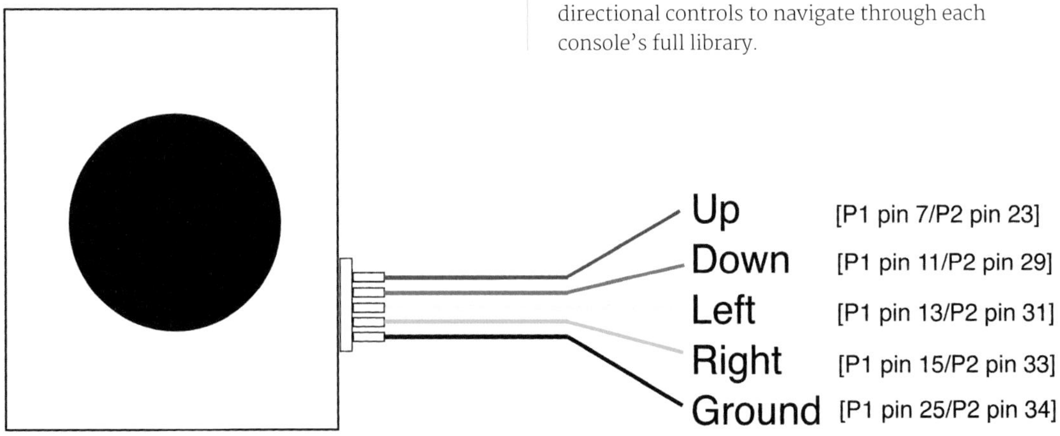

▶ Viewed from below, a standard Sanwa joystick's 5-pin connector goes to up, down, left, right, and ground. The diagram shows standard colour coding

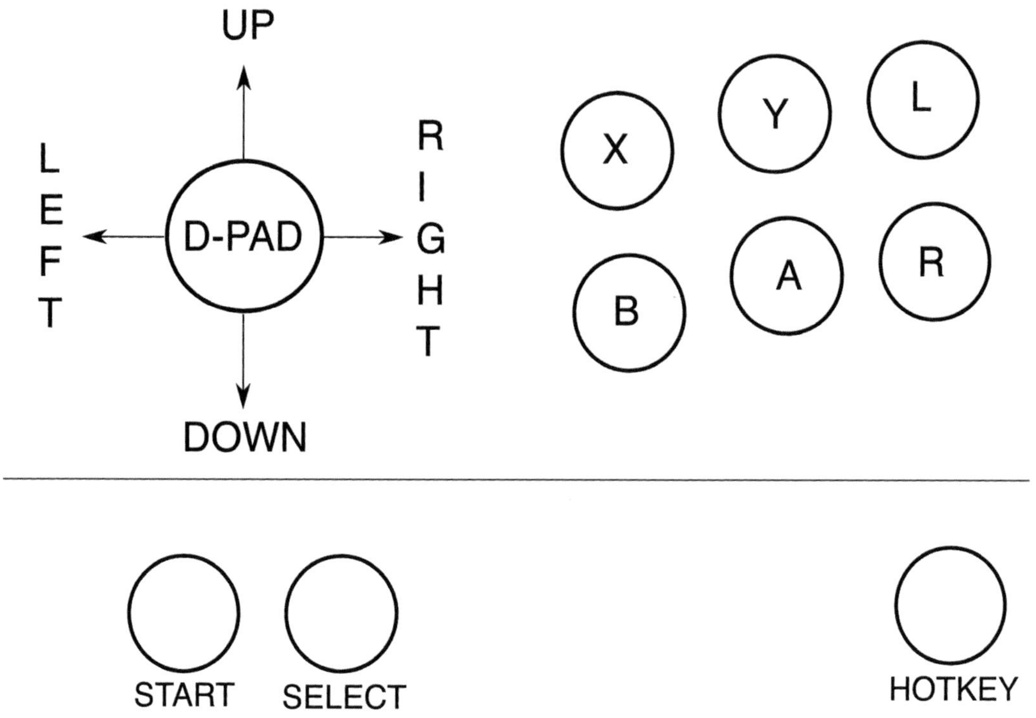

Button and joystick correspondences for player controls. The joystick maps to the D-pad. L and R correspond to L1 and R1, equivalent to the shoulder buttons of modern joypads

Warning!
Mains electricity & power tools

Be careful when handling projects with mains electricity. Insulate your cables and disconnect power before touching them. Also, be careful when using power tools during this build.

magpi.cc/drillsafety
magpi.cc/electricalsafety

09 recalbox.local

Once your arcade machine is connected to your local network, you'll be able to access it in a web browser via **http://recalbox.local**. On the main page, you'll see shortcuts to a virtual gamepad, keyboard, and touchpad, which allow you to navigate through the arcade machine's menus remotely.

To add some authenticity to older titles, go to Systems and set the Shader set to Retro, which will apply community-favourite shader and scanlines settings to each game. On the other hand, if performance is poor, disable shaders and rewinding here. Click Save at the bottom of the page to store your changes.

Below, the Configuration tab lets you set networking options, enable and disable the Kodi media player, and configure the behaviour of the EmulationStation front end and the hotkey.

10 Manage game & BIOS files

The easiest way to manage your game ROMs on Recalbox is via the web interface, where the ROMs tab lets you select the directory for your desired console, stop the EmulationStation front end, upload games, and restart EmulationStation to load them.

You can also copy games over to the **roms** directory in Recalbox's Samba share. Even if you don't plan on emulating a specific console, don't delete the containing folders for its games, as they're required.

Recalbox also shares a **bios** directory, where you can add freeware or legally purchased computer or console BIOS files.

11 Buy and install a game

ROMs and a functional BIOS set for a number of Neo Geo maker SNK Corporation's games are available to buy as part of the Neo Geo Classics Collection (**magpi.cc/neogeoclassics**).

You'll need a Windows, macOS, or Linux PC to install or extract these. You'll find the ROM and BIOS files in the install directory; for example, **ironclad.zip** and **neogeo.zip** respectively for the fantastic scrolling shoot-'em-up Ironclad. If you don't want the whole collection, you can buy Ironclad alone at **magpi.cc/ironclad**.

Connect to Recalbox via SMB and copy the game ROMs into **roms**, and **neogeo.zip** into **bios**.

Restart EmulationStation and you should find your new games in the Arcade game list. Not all of them will work out of the box. Start any of them and press the hotkey and B buttons to open the Libretro emulation interface. Scroll down and select Options > Neo-Geo mode > Use UNIBIOS Bios. We aren't using UniBios here, but the file supplied by SNK is compatible with this setting.

Top Tip

Preconfigured USB support

If your cabinet uses a USB controller board, then RetroPie won't need any extra drivers to detect your controls.

Your ef
Lv1
Lv1
INSERT COIN
1P-0011400
CREDITS 02
LEVEL-4
CREDIT 00

SNK has made plenty of its arcade ROMs available to buy. Ironclad for Neo Geo-based arcade machines is a particular favourite

Press A twice to go back and select Resume. Your game should start.

12 Tweak your games entries

To hide the games that come with Recalbox, from EmulationStation press Start > Main menu > Games settings > Hide preinstalled games. Unfortunately, you can't pick and choose which get hidden, but you can manually download and re-add any that you'd like to keep.

You can also disable the ports category by editing **recalbox.conf** to include:

```
emulationstation.collection.ports=1
emulationstation.collection.ports.hide=1
```

If you want to add images or change the titles of the games you've added to Recalbox, the easiest approach is to use the built-in scraper. Highlight the game in the menu, press Start > Edit game > Scrape. You can also add your own ratings and keywords in this menu.

13 Get more games

The creators of the MAME emulator have been given permission to distribute some early arcade games, which you can find for download at **magpi.cc/mameroms**.

Many other emulated arcade games have been released for use on modern computers, but some – including collections by SNK, Capcom, Irem, and Namco – require an additional extraction and re-bundling stage. You can find tools and game lists to help you buy and use these at RED-project (**magpi.cc/redproject**) and SF30ac-extractor (**magpi.cc/sf30ac**). Linux GOG users may also require innoextract (**magpi.cc/innoextract**). Non-Neo Geo arcade games should go into the **roms/MAME** directory.

The homebrew scenes for arcade games tend to focus on physical releases, but we've had luck with Codename: Blut Engel for Neo Geo and Santaball (**magpi.cc/neogeohomebrew**) for Neo Geo CD.

For more retro and homebrew games that work well with arcade controls, including Sega's Mega Drive Classics collection, see **magpi.cc/legalgameemu** and **magpi.cc/legalroms**.

Warning!
Copyright alert!

It is illegal to download copyrighted game or BIOS ROMs in the UK without the permission of the copyright holder. Only use official purchased or freeware ROMs that are offered for download with the consent of the rights holder.

magpi.cc/legalroms

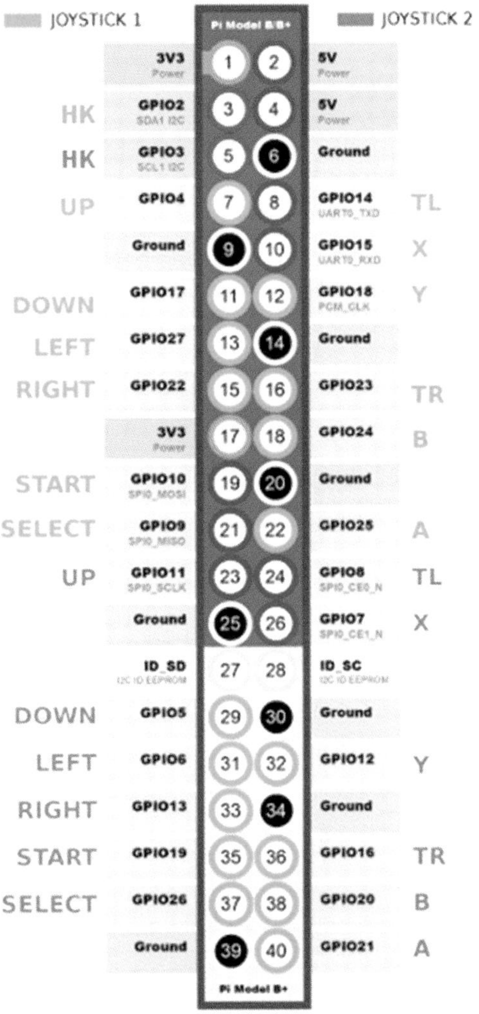

JOYSTICK 1
Pi Model B/B+
JOYSTICK 2
3V3 Power 1 2 5V Power
HK GPIO2 SDA1 I2C 3 4 5V Power
HK GPIO3 SCL1 I2C 5 6 Ground
UP GPIO4 7 8 GPIO14 UART0_TXD TL
Ground 9 10 GPIO15 UART0_RXD X
DOWN GPIO17 11 12 GPIO18 PCM_CLK Y
LEFT GPIO27 13 14 Ground
RIGHT GPIO22 15 16 GPIO23 TR
3V3 Power 17 18 GPIO24 B
START GPIO10 SPI0_MOSI 19 20 Ground
SELECT GPIO9 SPI0_MISO 21 22 GPIO25 A
UP GPIO11 SPI0_SCLK 23 24 GPIO8 SPI0_CE0_N TL
Ground 25 26 GPIO7 SPI0_CE1_N X
ID_SD I2C ID EEPROM 27 28 ID_SC I2C ID EEPROM
DOWN GPIO5 29 30 Ground
LEFT GPIO6 31 32 GPIO12 Y
RIGHT GPIO13 33 34 Ground
START GPIO19 35 36 GPIO16 TR
SELECT GPIO26 37 38 GPIO20 B
Ground 39 40 GPIO21 A
Pi Model B+

GPIO wiring: Connect your joysticks and buttons to Raspberry Pi's GPIO as shown. *Image by digitalLumberjack of the Recalbox project, licensed under GPL2*

PCBWay

Advanced PCB Fabrication

Instant quote online & User-friendly web interface
Fast turnaround in 24 hours
Perfectly implement your idea with different types of PCB!

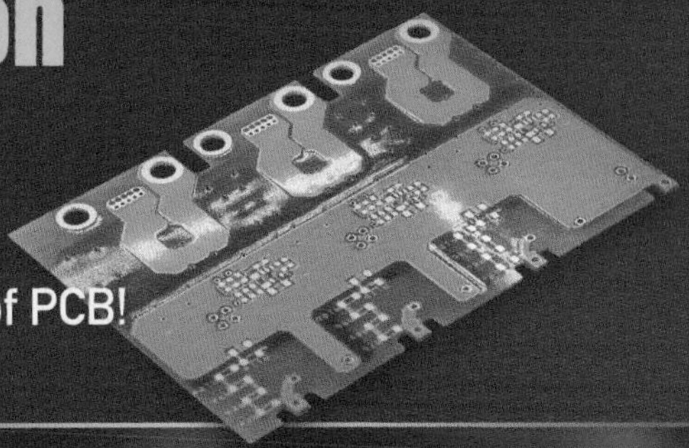

High-Quality PCB Assembly

Turnkey PCB Assembly Service
Start from only $30
Free stencil & Free shipping all over the world

15% off for Flex PCB

Professional Flex/Rigid-Flex PCB.
Quality assurance with the inspection and test report
Make high-end products the easy way

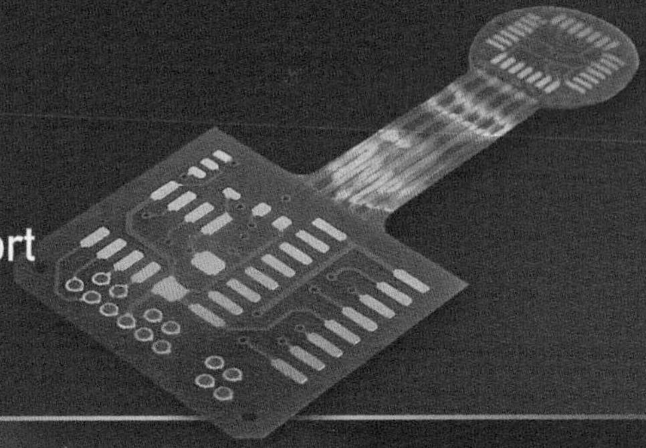

3D printing & CNC machining service is also available now!

More information:
www.PCBWay.com

Phone: (0571) 8531 7603
Email: service@pcbway.com

Build an arcade machine: Decorate your cabinet

You've built an arcade cabinet, but vinyl decals and edge moulding will bring it to life

MAKER

K.G. Orphanides

K.G. is a writer, maker of odd games, and software preservation enthusiast. Their household can now hold very retro Street Fighter II tournaments, and that's beautiful

@KGOrphanides

Most arcade cabinet kit suppliers print pre-designed or custom vinyl decals to decorate your cabinet. Third-party printers can produce vinyls to your specification, but make sure that you provide accurate measurements.

Our vinyl decals, bought from Omniretro (**magpi.cc/omniretro**), arrived on a roll and had to be cut out, but some firms will die-cut vinyls for you. We'll use a wet application process, which makes it easier to remove and reposition decals for a short while after initial placement, to help you get a perfect alignment.

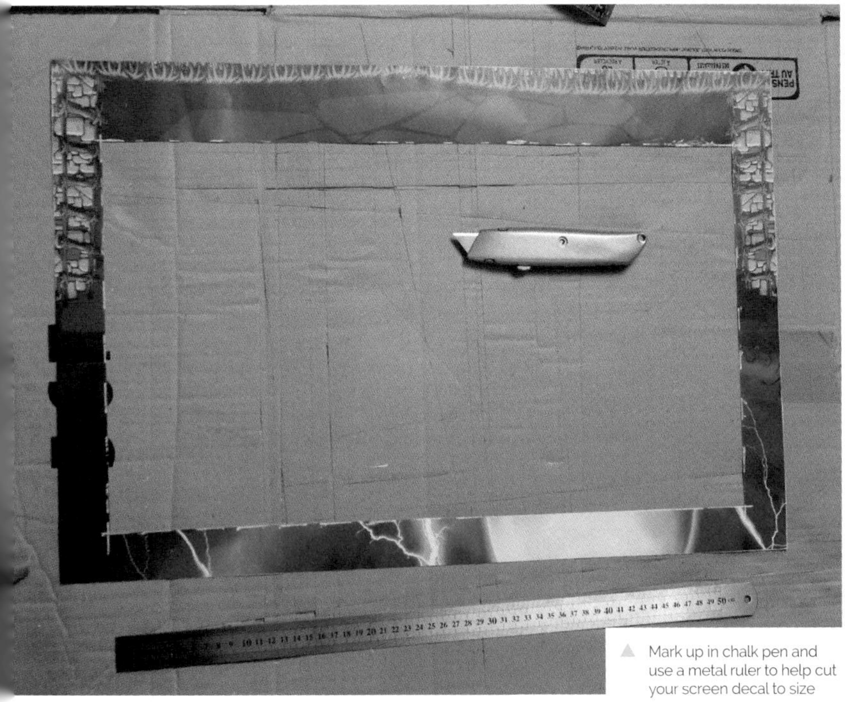

Mark up in chalk pen and use a metal ruler to help cut your screen decal to size

01 Flatten your vinyl decals

If your vinyls all came on a single roll, the first step is to cut each of them out. First separate them, if they're on a single roll, but leave generous margins. Spread them out on a table or on the floor and weigh them down – coffee table books and textbooks are good for this. Leave them for at least an hour or two: 24 hours is better.

02 Cutting out

Now they're flat, it's time to cut out your vinyls. Try to get rid of all white matter on straight edges. The easiest way is to line up a long metal ruler so that it just covers the edge of the printing, and run a scalpel down the outside of it. Curved sections for the cabinet side panels are trickier, but you don't need to worry about these as they're easy to trim down once fitted. For now, trim them freehand and leave as much white overmatter as you feel comfortable with.

03 Partial disassembly

Depending on the design of your cabinet, you may need to remove a side panel to take out the acrylic marquee and screen panels. Before doing this, use a liquid chalk pen and ruler to mark the edges of your LCD display on the acrylic, so we can accurately hide the bezel.

If you've previously fitted joysticks and buttons to your control panel, this is the time to remove them too. Apply steady pressure to the rear of snap-in style buttons to pop them out of the cabinet. People with large fingers may find a ButterCade Snap Out Tool useful for this.

04 Applying vinyl to your marquee acrylic

Two acrylic parts require individual application of vinyls: the marquee and the screen that goes in front of your monitor. The former is easy: remove the backing from the vinyl marquee decal and any protective film from the acrylic. Spray both the acrylic and the adhesive back of the vinyl with two or three squirts of application fluid. You want them to be damp all over but not awash.

Pick up the vinyl decal in both hands and, starting at one end of the acrylic, line it up with the edges and paste it down. If you're not happy with the positioning, firmly hold the vinyl and snap it back up – the application fluid will help it release easily.

Once it's positioned, use your applicator and a cloth to smooth it down, drive out any excess water, and remove any trapped air bubbles under the vinyl. Trim any excess vinyl spilling off the edge of the acrylic with a knife.

05 Measuring your screen acrylic

Cutting your screen decal to size is awkward. Before removing the screen acrylic from the cab, we marked the inner position of our monitor's bezel on the acrylic using a chalk pen. If your cabinet has a detachable VESA mount, bring the monitor with you to help line everything up.

> "Grab your screen vinyl and mark up the area to cut out"

Measure the distance between the edge of the acrylic and the chalk line you drew on it. Measure in multiple places to be sure of distances. Our 24-inch monitor's positioning and bezel size means that we cut 35 mm in at the top and sides, and 65 mm from the bottom – yours will differ.

06 Cutting your screen decal

Once you've taken the measurements, grab your screen vinyl and mark up the area to cut out. Mark on the side showing the picture, paying particular care to the corner positions. Double-check these by placing the acrylic on top to make sure both sets of marks line up.

The glowing logo is created using light-permeable vinyl on acrylic, with an LED strip mounted just behind it

A join between this bartop cabinet and its stand is rendered invisible by a large vinyl decal

Ghouls 'N Ghosts is available to buy in the Capcom Arcade Stadium collection on Steam **(magpi.cc/ghoulsnghosts)**

Use a vinyl applicator and a cloth to stick down, remove excess moisture, and eliminate air bubbles from your decals

Grab your metal ruler, place it along your marked line, and cut a rectangle out of the middle of the vinyl decal with a blade. If in doubt, err towards leaving too much vinyl rather than too little. To check positioning, put the acrylic over your monitor, and your vinyl over the acrylic: they should all line up.

You'll Need

- Vinyl decals
- U-moulding/T-moulding
- Scalpels/craft knives
- Strong scissors
- Liquid chalk marker pen
- Metal rulers, tape measures
- Vinyl application fluid
- Vinyl applicator
- Neoprene glue

07 Screen decal application

Now, turn the vinyl upside down, remove its backing, spray it and the acrylic with the application solution, and stick it down using an applicator and cloth. Residual chalk marks can be wiped off using a bit more of the application solution.

Allow both the marquee and screen decals to dry for a day, trim them if needed, slide them back into your cabinet, and reattach anything you removed. This will probably be the last time you do this, so make sure the side panels are on securely and are correctly lined up and bolted to your stand, if you have one.

If you plan on back-lighting your marquee, this is a good time to put in your light. We used adhesive tape and supplied clips to mount a 50 cm USB-powered LED light on the underside of the marquee, just in front of the speakers.

08 Applying flat vinyls

If you have a full-height cabinet or a bartop and stand, you'll probably have a number of flat, front-facing areas to decorate – in our case, the front cupboard door of our stand, its base, and the front of its foot. Do these next to get your hand in.

The drill is the same for all of them: place the vinyl decal face-down on the floor, remove its backing, spray both it and the surface you're

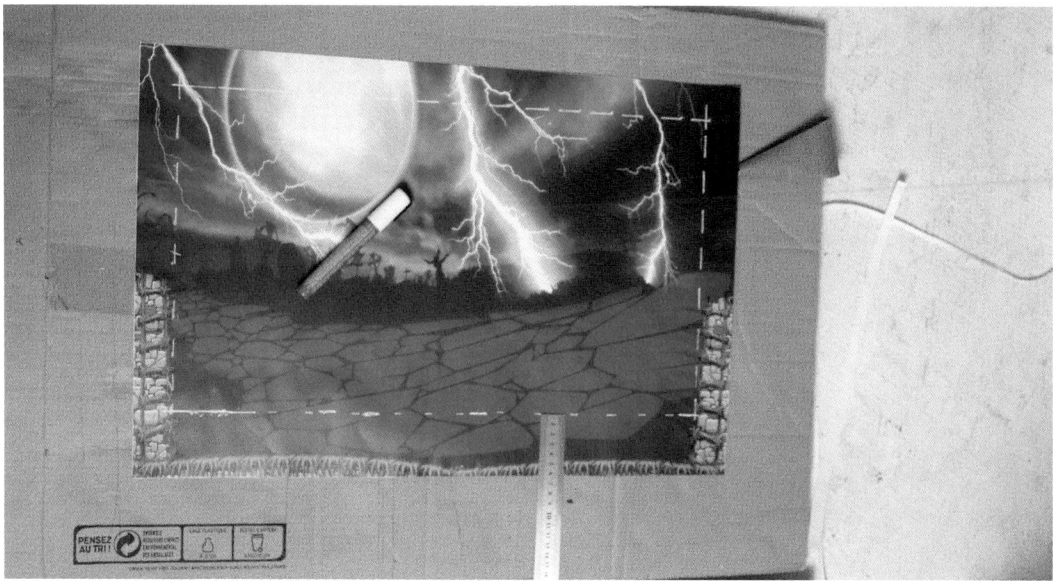

We marked the inner position of our monitor's bezel on the acrylic using a chalk pen

applying it to, position your decal, and smooth it out with your applicator. Use a scalpel to trim off any overmatter. For the door, we applied the decal with the door in place – knob removed, starting at the top. We had to open the door to flatten and trim the vinyl in places.

09 Control panel decals

Most control panel decals wrap around the top and front of your panel. Buttons and joysticks should not be present during application. This is a relatively easy section to apply, but watch your position if there are decorative patterns designed to surround specific buttons or joysticks.

You may need to trim overmatter from the sides with a scalpel to get the decal to fold over the front face properly. Be careful when smoothing the vinyl on this fold, as it can be prone to both trapped air bubbles and damage from the join beneath.

10 Cabinet positioning

Side panels are the largest pieces of vinyl you'll be applying, but they're less intimidating than they seem. For a standalone bartop, one person can mount them in a vertical position with little fuss, as shown in Omniretro's video at **magpi.cc/omniretrovinyl**.

Full-height cabinets present more of a challenge due to their height and the size of the vinyl - a second person is useful here. You can apply long vinyls in an upright position, but we'd already attached rubber feet to our cabinet, so we used these to help pivot the cab down to lie on a sheet of cardboard on the floor.

11 Apply side panel vinyls

Lying flat and sprayed down as before, it's easy to line up the side-panel decal. Make sure everything's covered – with two people, it's easy to snap the decal back up if you make a mistake, then use a cloth and applicator to drive out excess moisture. Use a Stanley knife to trim the vinyl to size – its solid metal body makes it easy to follow the line of the cabinet's curves.

Go around again to remove any air bubbles and ideally leave the vinyl to dry for at least a couple of hours before pivoting the cabinet back up and lowering it to expose the opposite side. Repeat the process.

▲ You can leave some white-space overmatter on side panel decals before application, as they're easy to trim with a knife afterwards

> Side panels are the largest pieces of vinyl you'll be applying

If your cabinet has separate stand and bartop parts, but uses a single sticker, there will be a slight ridge where these join. However, careful application (and a sympathetic vinyl design) makes this effectively invisible. Just be careful smoothing around it.

Top Tip

Screen materials

Acrylic scratches really easily, so tinted tempered glass is an excellent alternative for your cabinet screen.

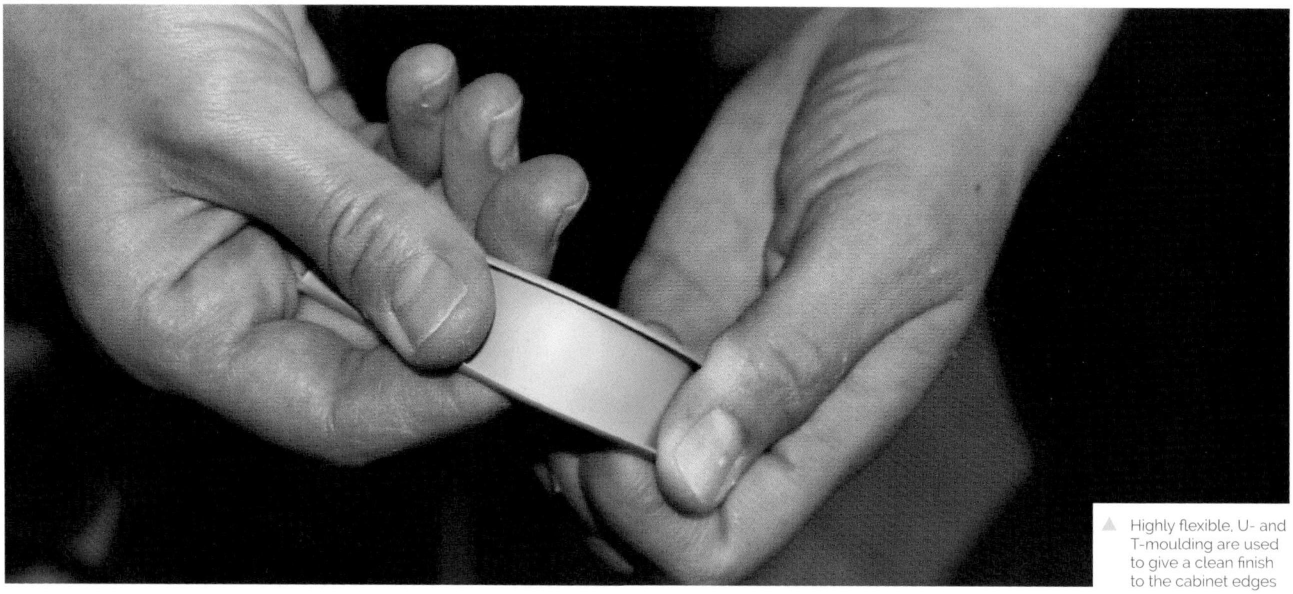

Highly flexible, U- and T-moulding are used to give a clean finish to the cabinet edges

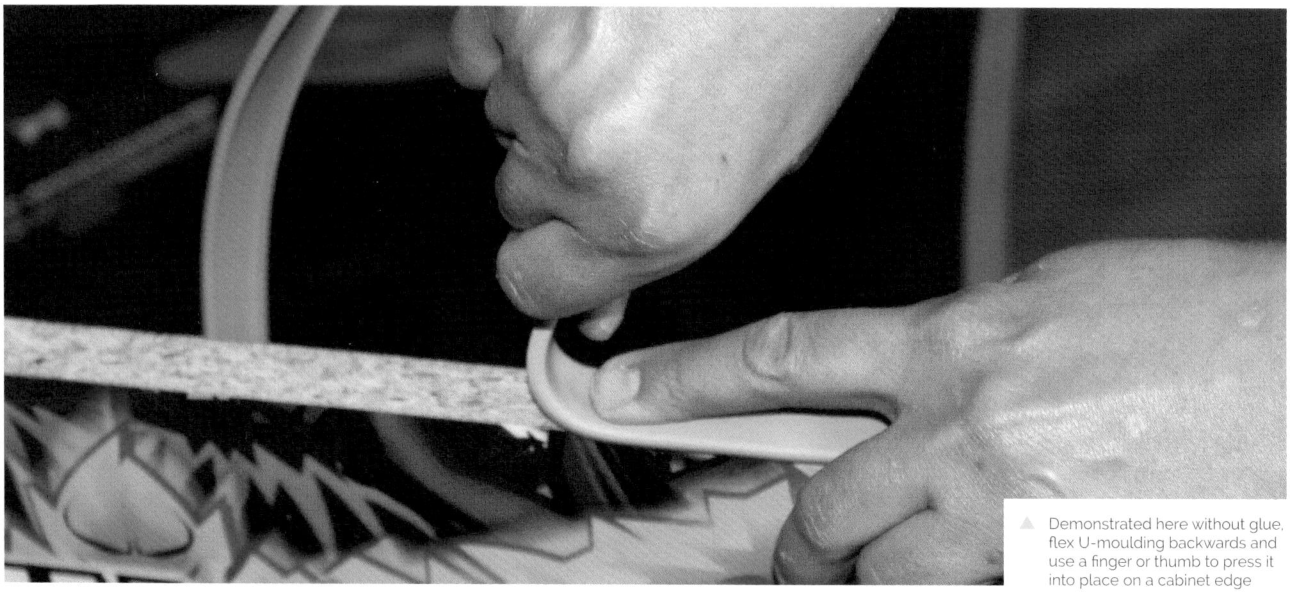

Demonstrated here without glue, flex U-moulding backwards and use a finger or thumb to press it into place on a cabinet edge

Warning!
Solvents

Always use solvents in a well-ventilated area and keep away from open flames.

magpi.cc/solvents

12 Moulding

We used U-moulding on our cabinet, with neoprene glue to hold it in place securely. First, measure and use scissors to cut two strips to go above and below the marquee – it's better to cut these a few millimetres too long and then trim than it is to have a gap.

Use a spatula to help apply neoprene glue along the edge you're working on, then use the tube's nozzle to apply glue to the inside of the U-moulding.

To lock U-moulding into place, bend it backwards to spread the U-shaped section, push that onto the edge you're applying it to, and then roll the moulding down along the edge, using a finger to push it into place.

When applying it to a long section, such as each side of your cabinet, start at the front underside – rubber feet help access here – apply glue to the cabinet edge and the first 50 cm of your roll of moulding, and have someone else feed it to you as you work up and around the cabinet. When you get to the bottom at the back, cut off your moulding with scissors.

T-moulding locks into a pre-cut groove along the edges of your cabinet, making it more secure, but it's still a good idea to apply glue to the flat surfaces for security. Either way, use a rubber mallet to gently tap down your moulding at the end.

You can use acetone to clean the glue off your hands and the moulding, but keep it away from the vinyl.

After spraying the vinyl decal, and the acrylic, with our homemade fluid, we applied it and smoothed down with an applicator and cloth

Vinyl application fluid

You can buy commercial vinyl application fluid (**magpi.cc/vinylfluid**), widely used by car customisation enthusiasts to apply decals, but we filled a spray bottle with the following homemade formula:

- 66 ml surgical spirit
- 132 ml water
- 2 drops washing up liquid

You can use warm water with a drop of washing up liquid alone, but the surgical spirit reduces drying times, which means less waiting between different stages of application and decorating.

13 Finishing moves

Use a scalpel to cut out the vinyl above the button holes: locate a hole, pierce it with the blade, slice until you find the edge of the hole, and then follow the hole round to remove all the vinyl. Do this for all your joystick and button holes.

As described in *The MagPi* #105 (**magpi.cc/105**), screw your joysticks back into place from the inside. If you're going to put protective acrylic panels over your control panel, this is the time to do it – they're held on solely by the buttons.

However, because our cabinet is for home use, we've left the vinyl bare for a more comfortable and attractive finish. If your cabinet will see lots of play, acrylic will protect it and cut down on wear and tear. Whichever you choose, connect a DuPont cable to each button and pop them into place.

Follow the instructions from issue 106 to connect your buttons and peripherals to Raspberry Pi.

Warning!
Sharp objects

Take care when using knives and scalpels.

magpi.cc/handknives

Build an arcade machine: RetroPie and stream from Steam

Use RetroPie as your arcade operating system and add extra emulators with support for Steam Link. Stream games from a powerful PC to Rasbperry Pi

MAKER

K.G. Orphanides

K.G. is a writer, maker of odd games, and software preservation enthusiast. Their household can now hold very retro Street Fighter II tournaments, and that's beautiful

@KGOrphanides

Last month, we used Recalbox for our main arcade cabinet operating system, but it's not your only choice. In this final instalment of the 'Build an arcade machine' series, we'll use the RetroPie distribution, currently at version 4.7, to provide extra features such as Steam Link support, as well as taking a longer look at where to buy arcade games and how to get them onto your system. This tutorial assumes that you already have a fully assembled and wired arcade cabinet.

01 Install and prepare RetroPie

Fire up Raspberry Pi Imager, connect your microSD card writer, and install RetroPie from its Choose OS menu. Re-mount the microSD card once you've finished flashing it, because we've got some changes to make.

CONFIGURING
GAMEPAD 36
HOLD ANY BUTTON TO SKIP
BUTTON A / EAST
BUTTON 7
BUTTON B / SOUTH
BUTTON 0
BUTTON X / NORTH
BUTTON 3
BUTTON Y / WEST
BUTTON 2
LEFT SHOULDER
BUTTON 4
OK

More button assignments are available on RetroPie than you have arcade controls. You can skip the ones that don't match up

As with our DB9 joystick project in issue 101 (**magpi.cc/101**), we have to tell the GPIO to treat the controls as pull-up switches. Recalbox, by comparison, implements this by default.

Create the **pullup.sh** file we've supplied (**magpi.cc/pullupfix**). You can put it anywhere you like – we stuck ours in **/home/pi/**. Now open **/etc/rc.local** on the SD card and, above the `exit` line, add:

```
/home/pi/pullup.sh
```

This will load your pull-up settings on boot. If you're setting up your disk on a Linux system, you can set **pullup.sh** as executable now. Otherwise, we'll do that on first boot.

02 First boot

Make sure you have a keyboard plugged into your cabinet for this bit. We left ours propped up against the marquee acrylic during setup for easy access. A Bluetooth keyboard is a viable alternative, but it's easier to start with a wired connection.

Plug in Raspberry Pi's power. It should boot to the EmulationStation interface, but we can't configure the controls until we've set our pull-up script executable. Press **F4** to exit to the command line and type:

```
chmod /home/pi/pullup.sh +x
```

While we're here, let's enable SSH:

```
sudo raspi-config
1 system option
s3 password
```

Raspberry Pi is powerful enough to run most games, and you can stream classic and modern arcade titles from a separate PC on your network via Steam Link

Steam Link lets you create a dedicated control map. Remember to map your hotkey button to its guide button

DOWNLOAD THE FULL CODE:

magpi.cc/rpipullupfix

```
enter a new password
3 interface options
enable ssh
yes
```

You can now SSH into Raspberry Pi from another PC using a client such as Remmina or PuTTY.

03 Add hotkey button support

If, like ours, your arcade cabinet's GPIO controller setup has either one or two extra hotkey buttons for easy access to save, load, and exit shortcuts while playing, then the standard version of the mk_arcade_joystick_rpi driver available from RetroPie's package manager won't support them. We'll have to manually add an updated version from maintainer Recalbox's GitLab repo.

At the command line, type:

```
git clone --branch v0.1.9 https://gitlab.
com/recalbox/mk_arcade_joystick_rpi.git
sudo mkdir /usr/src/mk_arcade_joystick_rpi-
0.1.9/
cd mk_arcade_joystick_rpi/
sudo cp -a * /usr/src/mk_arcade_joystick_
rpi-0.1.9/
nano /usr/src/mk_arcade_joystick_rpi-0.1.9/
dkms.conf
```

In this file, change `PACKAGE_VERSION="$MKVERSION"` to `PACKAGE_VERSION="0.1.9"`. Press **CTRL+X** to exit, then **Y** to save.

Back at the command line, enter:

```
sudo dkms build -m mk_arcade_joystick_rpi
-v 0.1.9
sudo dkms install -m mk_arcade_joystick_rpi
-v 0.1.9

reboot
```

> We'll use the RetroPie distribution to provide extra features such as Steam Link support

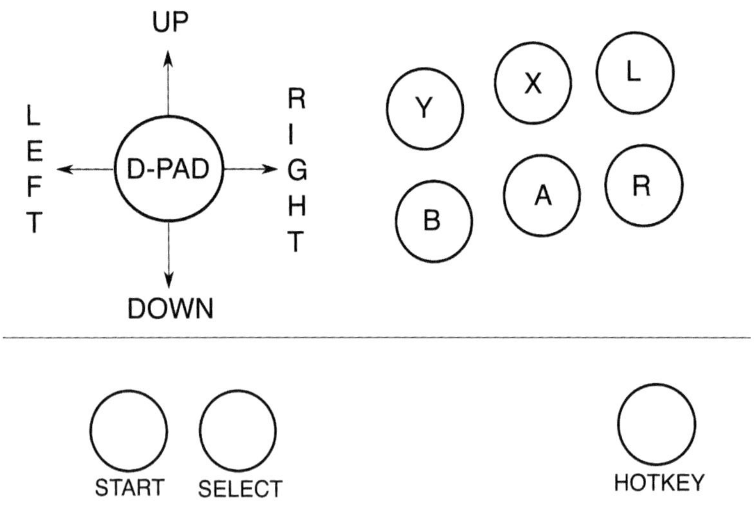

Button and joystick correspondences for player controls; we recommend this configuration for use with RetroPie. L and R map to the right and left shoulder buttons

04 Optional: Load your hotkey driver

When Raspberry Pi has rebooted, SSH back in and type:

```
sudo modprobe mk_arcade_joystick_rpi map=1,2
```

Go over to the arcade machine and press **F4** to get to the command line and test your controllers:

```
jstest /dev/input/js0
jstest /dev/input/js1
```

If that works, it's time to load that module on boot. At the command line:

```
sudo nano /etc/modules
```

In this file, add the following on a new line, then save and exit.

```
mk_arcade_joystick_rpi
```

Next, at the command line:

```
sudo nano /etc/modprobe.d/mk_arcade_
joystick.conf
```

In this file, add the following:

```
options mk_arcade_joystick_rpi map=1,2
```

Now save, exit and reboot.

05 Configure RetroPie

There's a bit more configuration to do before RetroPie is ready to go. SSH in and type:

```
sudo ~/RetroPie-Setup/retropie_setup.sh
```

...to open the ncurses configuration menu.

If you did not manually install a hotkey version of the mk_arcade_joystick driver in the previous steps and do not need one, go to:

```
P manage packages
driver
819 mkarcadejoystick
```

...and install it.

If you need to connect any Bluetooth keyboards or controllers, go to:

```
C Configuration / tools
804 bluetooth
```

Press **R** to register a device and follow the pairing instructions.

832 samba in the configuration menu sets up Samba shares so you can easily transfer ROMs and BIOS images over your local network

You can add extra emulators here, but we'll come to that later. For now, select the R Perform reboot option from the main menu.

06 Input configuration

When RetroPie reboots, it should inform you that it can detect two GPIO controllers. Press and hold any button on the left-hand button bank to configure controls for player 1. Because arcade controls don't map perfectly to a gamepad, you'll have to skip some buttons by pressing and holding any key.

> "When RetroPie reboots, it should inform you that it can detect two GPIO controllers"

Map up, down, left, and right on the arcade stick to the D-pad. Follow our button assignment diagram to map the top row to button Y, X, and L(eft shoulder), and the button below to buttons B, A, R(ight shoulder).

Map Start to player 1's left-hand front function button and Select to their right-hand front function button – this will be their 'insert coin'

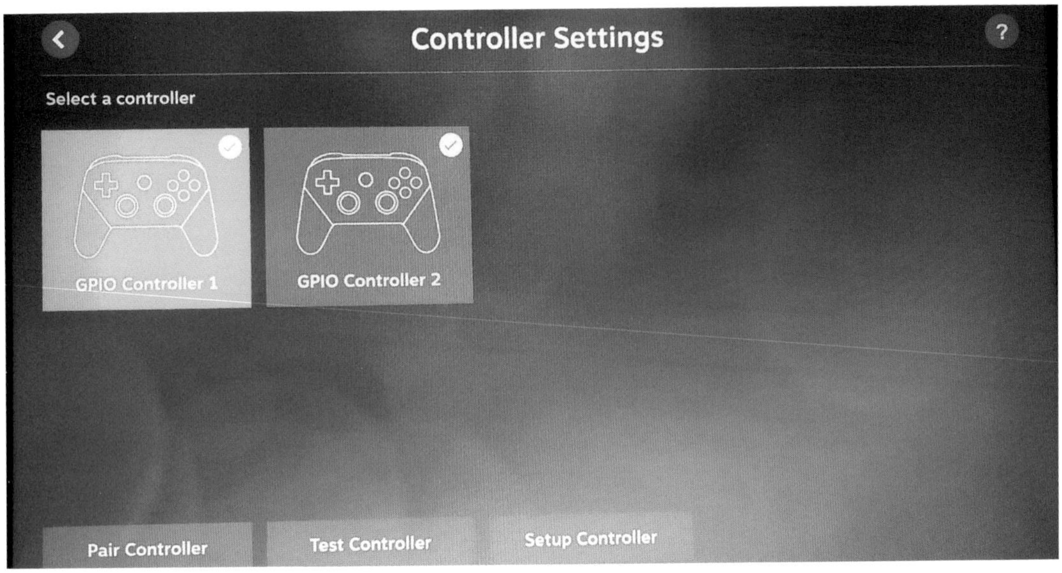

◀ You'll probably want to reconfigure your controls in Steam Link to better match its Steam Controller-based expectations

button. In our wiring configuration, our single hotkey button – the last we set – is associated with player 1.

Approve your configuration, then set up player 2's controls in the same way.

07 Getting to know RetroPie

With your controllers configured, RetroPie's main interface will open. Press A to select menus and items and B to go back. Press Start to open the main menu and Select to open the options menu. Press the same button again to close each of these.

As you have yet to put any games on the system, only the RetroPie menu will be available. Here, you'll find easy access links to configuration tools, including some we used earlier. Install new emulators and drivers from the RetroPie Setup menu.

You'll probably need to disable overscan to get rid of a black border around the screen. In the ReotroPie menu, select Raspi-config > Display options > Underscan > No and then reboot to solve the problem. Note that button B is mapped to the **ENTER** key in this set of menus.

When you add any new games, ROMs or emulators, you'll have to restart EmulationStation by pressing Select, going to Quit, and then Restart EmulationStation.

08 Install more emulators

Although this is an arcade machine, you can play what you like on it. The core lr-mame2003 and lr-fbneo emulators are included, along with those for popular consoles such as the Sega Mega Drive, used in some arcade systems and for which original games are legally available.

Some emulators require system BIOS images. Sadly, very few of these have been made legally available to emulation enthusiasts. SNK distributes a UniBIOS compatible BIOS set in its 40th Anniversary Collection. We recommend adding the following:

opt > 327 opentyrian – arcade-like DOS shoot-'em-up Tyrian 2.1 is now freeware.

exp > 241 lr-mame – a more up-to-date version of MAME that supports a wider range of ROMs. Install from source for bleeding edge.

exp > 307 digger – a sanctioned remaster of Windmill Software's Dig-Dug.

exp > 334 steamlink – this allows you to stream less emulation-friendly titles directly from a Steam installation on a Windows or Linux PC.

09 Configure your emulator

Once you've installed a new emulator, such as lr-mame, you'll have to configure the libretro back end to use it by default for either all games or selected titles. The easiest way to do this is to browse to the game you want to play in the EmulationStation front end.

Go to the Arcade menu, press B to start any game – it doesn't matter if it currently works or not – and then press B again when you're briefly prompted to 'press a button to configure'. Select option 1 to set the default emulator for arcade

Top Tip

Steam Link smoothly

Use a wired Ethernet connection for optimal Steam Link game streaming.

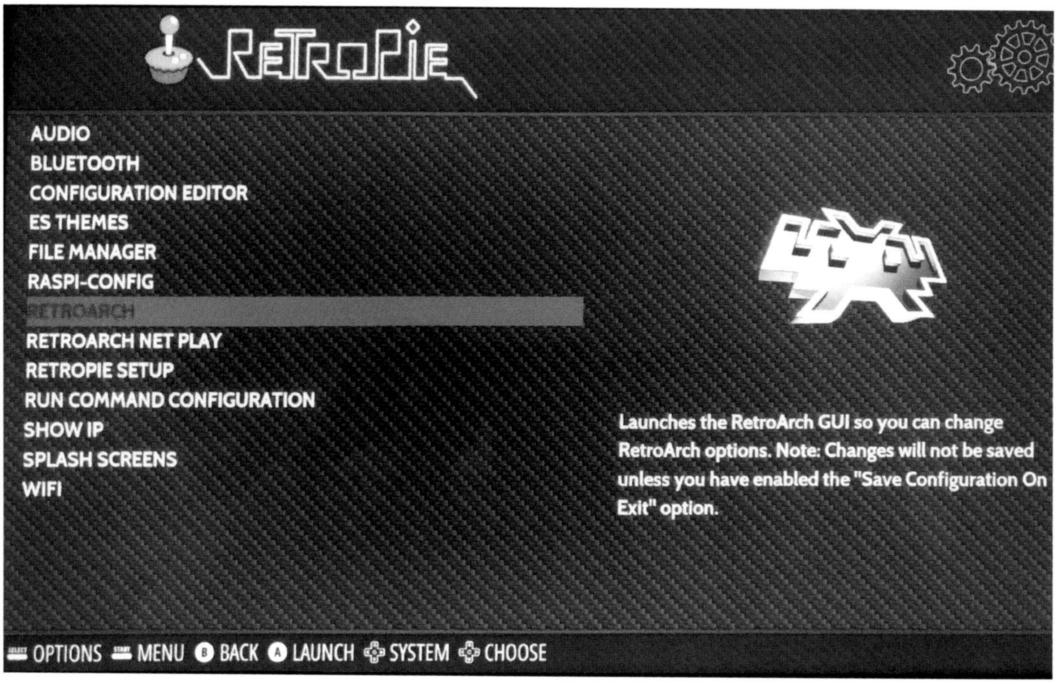

RetroPie is significantly more configurable than Recalbox, although its interface doesn't look quite as slick

Top Tip

A screw loose

If the ball on your joystick is loose, use a screwdriver in the slot on the underside of the stick or cloth-wrapped pliers to hold the shaft still while you tighten the ball.

games and choose lr-mame. Option 2 allows you to select a different emulator for anything that doesn't work well with this.

10 Connect Steam Link

Linking Steam to your arcade cabinet lets you stream a wealth of modern and classic arcade games to Raspberry Pi from a more powerful PC, like Melty Blood, Guilty Gear, Horizon Chase Turbo, and Street Fighter V. After you've installed it and restarted EmulationStation, go to the Ports menu and select Steam Link.

It'll download updates – you will need a keyboard plugged in to approve these – and then run. Make sure Steam is running on a PC on your local network and that Enable Remote Play is ticked under Settings > Remote Play.

On the arcade machine, select the computer you want to link to. Steam Link will show a code. Enter this in your PC's Steam client when prompted. To avoid a resolution mismatch, run Steam with a monitor that matches the resolution of your arcade machine set as your primary display.

11 Configure Steam Link

You may want to reconfigure your controls, as Steam Link doesn't inherit the control layout from RetroPie's EmulationStation, and some games do better with alternative button assignments – for example, to more closely match an Xbox or Steam Controller, which swaps the position of the B and A buttons.

To set these, launch Steam Link, press up to highlight the gear icon, press A (per our button assignment diagram), go right to highlight Controller and press A. Select the controller you wish to configure, then press down and right twice, and select Setup Controller.

Hit the button you want to associate with each Steam Controller-style button as it's displayed on screen. Use a keyboard or your second set of controls to use the skip button at the bottom to bypass extraneous buttons.

12 Why use Steam Link?

Steam Link is an invaluable tool for arcade emulation enthusiasts, not only because you can play more CPU-intensive games, but also because it's the best way of ensuring copyright compliance for a number of re-released arcade games.

We've been playing Ghouls 'N Ghosts from the Capcom Arcade Stadium on our cabinet via Steam Link. Unlike some SNK and Sega re-releases, Capcom doesn't supply emulator-ready ROM files and the EULA for that compilation doesn't allow you to extract its PAK files.

Neil Brown of decoded.legal opines (**magpi.cc/romextractionlegal**) that "when even a legitimate Steam purchaser extracts the ROMs and runs them on their own Raspberry Pi, they infringe Capcom's copyright", making streaming these titles your best option for fully legal home arcade action.

MIDAS
Displays
design • manufacture • supply

Plug & Play Raspberry-Pi HDMI TFT Displays

Midas first introduced the HDMI TFT range in 2018, and have continued to develop this range over the past 4 years. These displays are ideal 'plug & play' devices for a Raspberry Pi SBC.

With the addition of a Raspberry Pi SBC, the Midas Displays HDMI TFT modules become a fully functional, easy-to-use displays. The integrated PCB has a 40-pin GPIO interface and a standard 19-pin type-A HDMI connector, enabling direct connection of a Raspberry Pi SBC. The HDMI TFT modules can also easily be used with many other single board computers. These key features enable the HDMI TFT displays to be used with any external HDMI input source.

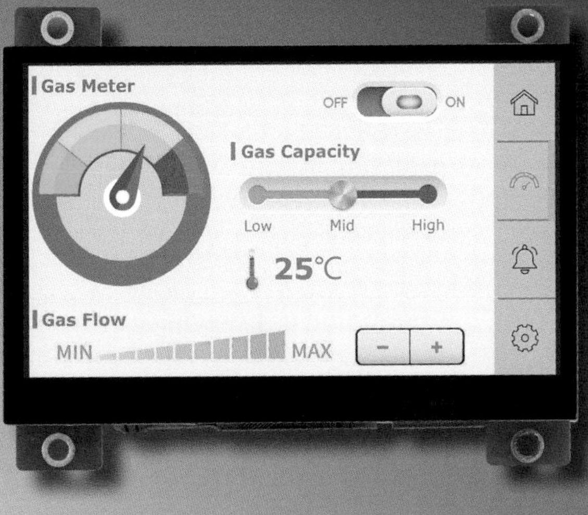

Key Features

- 4.3", 5.0", 7.0", 10.1", & 5.2" bar-type, sizes available
- IPS all-round viewing options
- 'Plug & play' with Raspberry Pi
- HDMI interface
- Capacitive, resistive, and non, touch options available
- Up to 1100cd/m ^{2} brightness

The products within the Midas HDMI TFT range are available to order direct, and through our distribution partners: Farnell, RS Components, and Digi-Key Electronics.

www.midasdisplays.com
+44 (0)1493 602602
sales@midasdisplays.com

Turn Keybow 2040 into a stream deck

Reprogram the RP2040 auxilliary keyboard into a powerful macro keyboard for streaming and more

MAKER

Rob Zwetsloot

Rob is amazing. He's also the Features Editor of *The MagPi*, a hobbyist maker, cosplayer, comic book writer, and extremely modest.

magpi.cc

You'll Need

- Keybow 2040 **magpi.cc/keybow2040**
- OBS **obsproject.com**
- VoiceMod **voicemod.net**
- USB-C to USB-A cable

Streaming is fun. Whether you're doing art, playing games, or just chatting with followers, it can be great to interact with folks while doing something else.

It is also quite complex to stream. Starting with a complicated OBS setup to stream your game, and ending up with loads of extra tabs and audio mixers thrown in, there can be a lot to control and maintain while you're live, and even more to do if you want to use silly audio effects or zooms, etc. This is where a stream deck comes in, allowing for short cuts for various functions at the press of a button. A lot of dedicated stream deck products can be pretty expensive, though, so let's make something a lot cheaper and far more customisable.

01 Construct Keybow 2040

The Keybow requires some building before you use it – it's fairly simple though. You attach some standoffs, a middle layer, add the switches to the final plate, add the keycaps, and then sandwich it all together. There are some illustrated instructions here: **magpi.cc/keybowbuild**. You should be done before an episode of *The Simpsons* is over.

For stream deck use, we also have a 3D-printed switch stand that it fits onto nicely, keeping the keys in eyesight (which is important, as we'll explain later).

02 Basic Keybow 2040 usage

Plugging your newly built Keybow 2040 into your PC will activate the pre-installed code that turns it into a numpad using a basic HID (human interface device) script. The LEDs in the keys are turned off, but go green as they are pressed.

As well as being usable while plugged in, you can also access the storage inside and find different examples in the **examples** folder. To use one, merely copy it to the top Keybow directory and change the file's name to **code.py**.

> A lot of dedicated stream deck products can be pretty expensive

03 Customise your code

The **code.py** file is what tells the Keybow 2040 how to operate. Looking through the examples, you'll be able to see how a lot of the individual parts work together. A full list of the functions can be found on the Keybow GitHub here: **magpi.cc/keybowgit**.

As it runs on CircuitPython, based on MicroPython, a lot of Python knowledge and skills are transferable while writing any custom code. With one caveat though – using sleep is not recommended by Pimoroni, and it did not work with this build either. Still, there's a lot you can do.

Make quick changes that affect your stream without having to fumble with all the open applications

Control your streaming with hotkeys activated from the stream deck

It's very quick to prototype and test your custom code as well – opening, editing, and then saving the **code.py** file will immediately reset Keybow 2040 so that you can test out your handiwork.

04 Keyboard shortcuts

The basic keys used by the default code should be altered if you plan to use this as a stream deck. They function as a numpad, and while they will work as hotkeys, they may also be used normally by the software you're running. Luckily, the Arduino key library allows for multiple keys to be sent at once:

```
keyboard.send(Keycode.LEFT_CONTROL, Keycode.
KEYPAD_ZERO)
```

In our code, we're still using numpad keys, but we've added the left **CTRL** key to the combo. You could easily use a different key, such as the scroll lock, or multiple keys. Just check to make sure the key combos you're planning on using aren't being used in other programs – we don't want any **ALT+F4** incidents while streaming games.

05 LED colours

We've opted for all the keys to be lit up while the stream deck is connected, and this can be done using a simple `keybow.set_all()` command, where an RGB code is placed in the brackets.

For our code we're using green (0, 255, 0), yellow (255, 255, 0), and red (255, 0, 0), and you can easily add or modify the colours as you see fit. All the keys light up green when the Keybow is on (mostly to remind us that it is), and keys will turn yellow once activated. The red is used for the mute and voice change button when activated, just to differentiate them.

For other colours, blue is a good one (0, 0, 255), as well as white (255, 255, 255). If you need another shade, simply enter the name of it followed by 'rgb' into your search engine of choice.

06 When pressed

One of two most important parts of our code is the `press_handler` function. As the code cycles, it waits for a key to be depressed before it starts doing anything. It gets told which key is being depressed, finds out which part of the key

Top Tip

Manual combos

The key combos you've set up can also still be done on your regular keyboard, in case you need a backup way to use a hotkey.

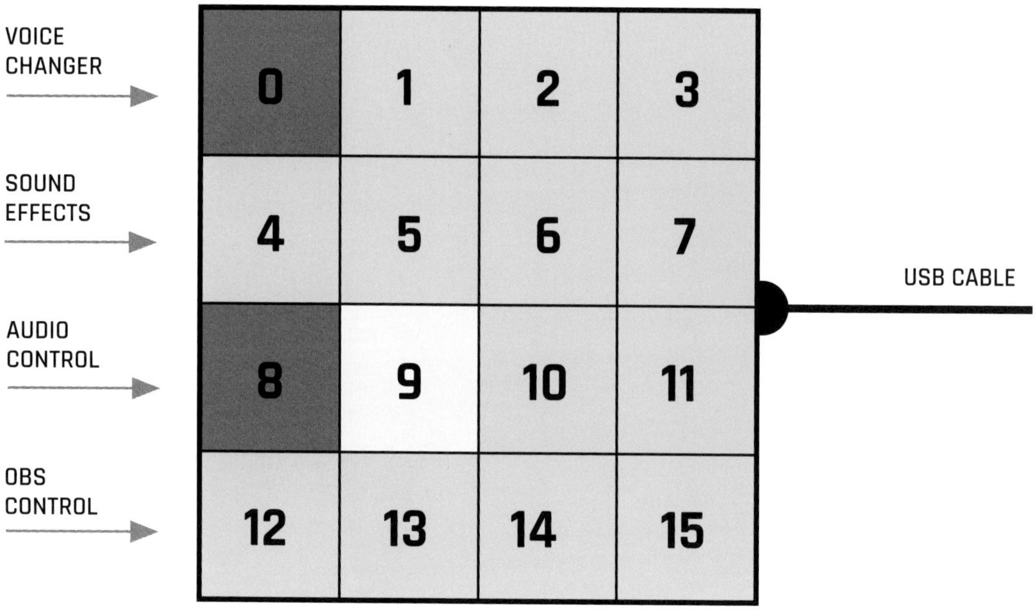

Here's how the keys are laid out on Keybow 2040 related to the code

Top Tip

OBS focus

You may find some difficulties using hotkeys with OBS – make sure the window is in focus for them.

combo it should use, and then sends that as a key press.

We also have an `if` statement so that a specific key turns to yellow when depressed, as all this button does is stop any sound effects from playing – it can easily be applied to other keys by modifying this `if` statement.

07 When released

By having separate press and release functions, we save the code treating a button-press as multiple events. We can also define what happens when the button is released; in this case we change the colour of the key to indicate that it's been used, as well as updating the `keytoggle` list, which we use to remember the state of the key (on or off, 1 or 0)

As we have many different buttons doing different things, we've racked up a long `if` statement – most are simply 'change between two colours'; however, one of them treats a series of three keys differently: `elif key.number >= 1 and key.number <= 3`. Only one may be lit yellow at a time, and we'll be using this to select a type of voice modification.

The `and` here also works as a Boolean operator, where AND means only if it satisfies both requirements, and OR means if it satisfies at least one requirement.

08 Planning your layout

For the code in this tutorial, we've set up the keys in categories, as shown in **Figure 1**. Keys 0–3 on the top line handle voice changing, the second row we use for sound effects, the third line for muting and stopping sound effects, and the last line for OBS hotkeys.

Remember, the keys go from 0 to 15 only when the Keybow is orientated with the USB cable sticking out of the right side.

Depending on your stream deck uses, you can make your own layout; however, be aware of the placement when programming.

09 Special keys

Our mute key and voice change key both turn red when activated. These are both important keys in the setup and if you're a bit forgetful, like us, you may not realise that the voice change key is still on until a helpful member in chat points out that you sound like a robot.

They broadly function the same as any other key pressed; however, the differing colours do help

Adding hotkeys to OBS is very easy, and it can control just about every aspect

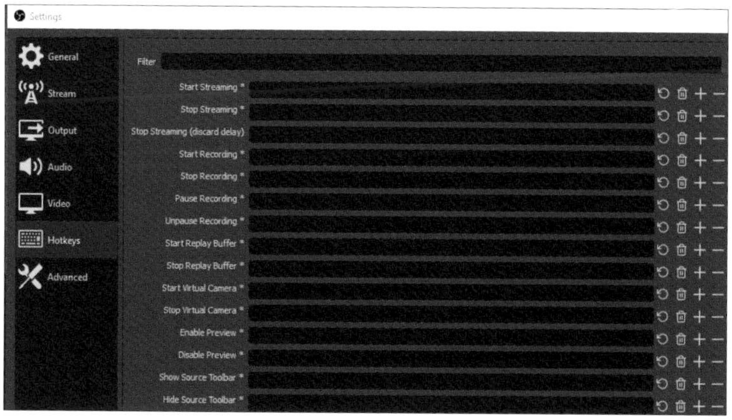

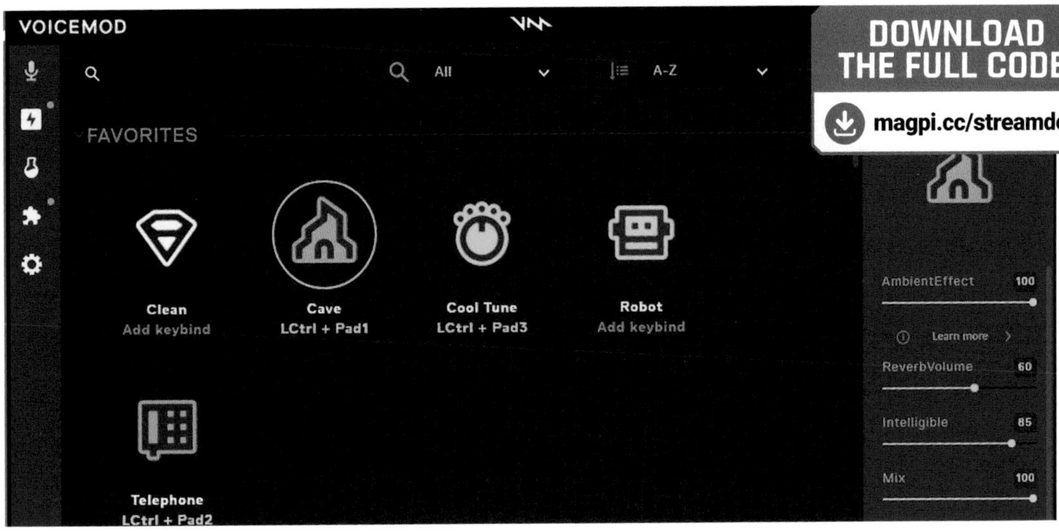

DOWNLOAD THE FULL CODE:

magpi.cc/streamdeck

The free version of VoiceMod offers many features, and they all work with hotkeys

when you're trying to quickly bump a key while drifting a Ferrari around LA.

10 Toggle keys

The keys 1, 2, 3 we are treating as toggle keys, as explained above. Here is how the code works:

```
key.set_led(*rgb2)
keytoggle[key.number] = 1
```

This part sets the key to yellow, and changes the `keytoggle` list to on/1 for that key.

```
if key.number != 1:
    keys[1].set_led(*rgb1)
    keytoggle[1] = 0
```

This part checks to see if key 1 was not pressed, and the following two statements check for keys 2 and 3. If those keys were not pressed, the colour is returned to green, and the `keytoggle` is set to off/0.

> "Our mute key and voice change key both turn red when activated"

11 OBS hotkeys

Just about every element in your OBS setup will be able to have a hotkey attached – whether it's simply changing to a scene, showing and hiding an image, or to stop streaming. From OBS, click Settings and then select the Hotkeys tab.

Click on the bar next to the function you'd like to hotkey, and then press the desired Keybow key. It will not only set the hotkey but also tell you what it is – a good way to check that your code worked properly.

12 VoiceMod hotkeys

VoiceMod is a great piece of software for doing fun voice changers, playing sound effects, and generally controlling the entire audio input for your stream. Selecting a modifier, you'll see a button to add a hotkey – click that and use the desired Keyboy 2040 key. The same is true for any sound effects or clips you upload.

In the Settings, you can also add other hotkeys – for the code we've used, key 0 is for voice changer toggle, key 8 for mute/unmute, and key 9 for stopping any sounds.

Elevating and angling a stream deck can make it easier to press buttons... and remember it's there

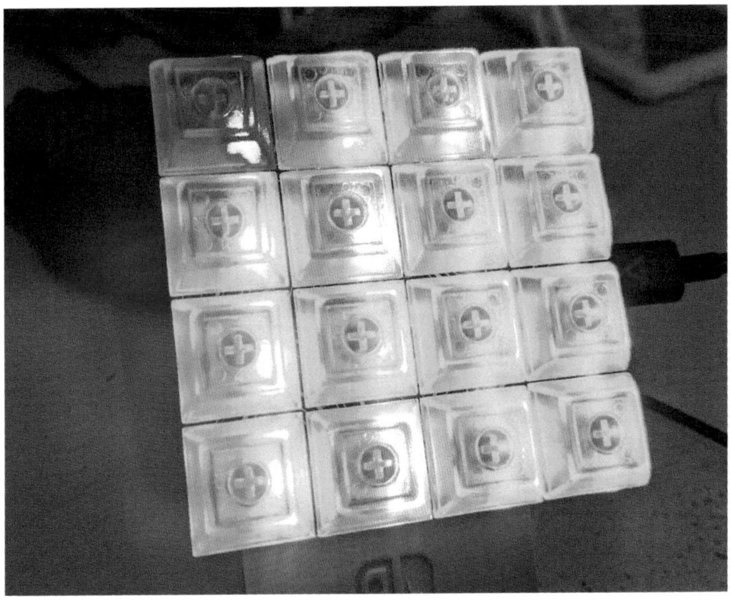

MAKE GAMES WITH RASPBERRY PI

Did you know that Raspberry Pi is a game-creation machine? There are many ways to write games and **Mark Vanstone** will show you a few to get you started

MAKER

Mark Vanstone

Educational games author from the 1990s, author of the ArcVenture series, disappeared into the corporate software wasteland. Rescued by Raspberry Pi!

magpi.cc/technovisual

The UK computer games industry has grown and grown since its origin in the eighties; grown so much that it's on a par with the UK film industry. This trend is true in many other areas of the world. If you are learning how to write games, then Raspberry Pi is a great way to get your teeth into the subject. There are many ways to write games and get started quickly.

If you are just starting out and not ready for text-based coding, don't worry. There are block-based systems like Scratch where you can lay out your game graphics on the screen. You can code games using blocks that you drag and drop into place to create your program. Once you have mastered that, you may want to move on to text-based coding like Python and Pygame. If you aim to start a career in the games industry, you will find that these days game designers use both methods: visual block editing and text-based programming. One of the most popular game engines, Unreal Engine uses a block editor called Blueprint that is underpinned by libraries of C++ code.

Game programming is a great way to learn a wide range of techniques that are useful for other areas of programming. You don't need to start with advanced scripting but can easily get quick results with the tools in this article. Let's make some games!

MAKE GAMES WITH SCRATCH

Scratch is an ideal place to start making games, with a great online community of creators who share their games so that you can see how they were made

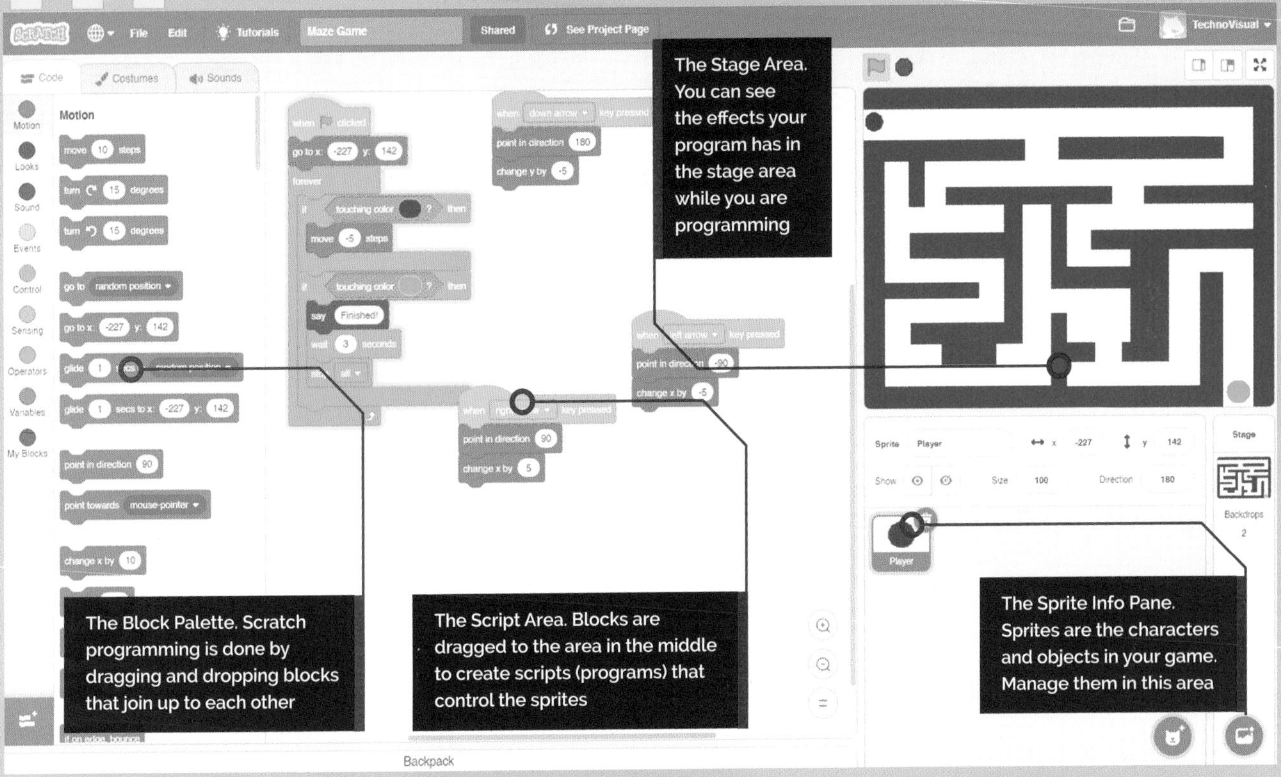
The Stage Area. You can see the effects your program has in the stage area while you are programming
The Block Palette. Scratch programming is done by dragging and dropping blocks that join up to each other
The Script Area. Blocks are dragged to the area in the middle to create scripts (programs) that control the sprites
The Sprite Info Pane. Sprites are the characters and objects in your game. Manage them in this area

Top Tip

Using sounds

If you want to use sound in your project, you can go to the Sounds tab. You will find a cat meow sound to start you off.

Scratch is a block-based visual editing programming language. Instead of writing commands in text, you click and drag objects (known as 'sprites') and control them with block commands. It's designed to make object-oriented programming easy to understand, and is a great way to get to grips with coding concepts. Due to its visual nature, it's ideal for creating basic games and interactive stories.

There are several versions of Scratch that are compatible with all versions of Raspberry Pi although the latest version, Scratch 3 is recommended for Raspberry Pi 3 and Raspberry Pi 4.

01 Get Raspberry Pi ready

It's always a good idea to keep your system files up-to-date. You can either download a fresh install of Raspberry Pi OS to your system card using the instructions at **magpi.cc/imager** or from your Terminal window use the commands `sudo apt update` and then `sudo apt upgrade`. It is wise to always go through this procedure before installing anything new on your Raspberry Pi to make sure you have the latest version of all the system files. Of course, for any installs or updates, you will need a connection to the internet.

02 Install Scratch

There are three versions of Scratch and an online editor. You can install Scratch 3 by clicking on Menu > Preferences > Recommended Software. In the Programming section, you will see Scratch 3. Place a tick in the Install checkbox to the right and click Apply.

03 Your first Scratch

If you have not used Scratch before, you probably want to jump straight in and make something happen. With Scratch, you can do just that. You'll find Scratch in Menu > Programming > Scratch 3. You will see a cartoon cat on the right-hand side and a set of blue boxes on the left. Drag the **turn 15 degrees** block into the Script area in the middle (this is where you assemble your program). Click the **turn 15 degrees** block and the cat will rotate.

04 The green flag

Our rotation block is good, but what if you want something more? We can build a program of blocks by joining them together. Click Control in the sidebar and then drag and drop the **repeat 10 block** into the Script area. Then move your **turn 15 degrees** block so that it's inside the **repeat** block. Then click Events in the Blocks palette and drag the **when (green flag) clicked** block to sit on top of the **repeat** block. Now if you click the green flag at the top of the window, the program will run.

05 Customising your sprites

In Scratch, a sprite is a graphic on the screen that you are controlling. You can change what it looks like by loading in your graphics or editing the image. Select the Costumes tab at the top of the screen. From there you can use the painting tools to make a new image or alter the one that is already there. Try drawing some lines or filling in some shapes to see how it works. You can also write text. If you select items with the arrow tool, you can also change their colour.

RASPBERRY PI SCRATCH PROJECTS

Go to the projects section of Raspberry Pi's website, **magpi.cc/projects**. In the 'Find a Project' section, select Games from the Topic drop-down, and Scratch from the Software drop-down. You will be shown a selection of game projects for Scratch. Each of these projects is laid out as a step-by-step tutorial to help you build the game. There are lots of different game projects available, so you shouldn't run out!

Or you can try PICO-8

PICO-8 is a fantasy console for making, sharing, and playing tiny games and other computer programs. It feels like a regular console and runs on a variety of platforms. It has a suite of cartridge creation tools and an online cartridge browser called SPLORE. The programs are distributed in the form of a PNG file and each program has a memory limit of 32kB, so it is like programming a retro-style, 8-bit computer. The PICO-8 development system costs £11 / $15 and can be downloaded from **magpi.cc/pico8**.

Top Tip

Loading and saving

You can load/save Scratch projects to your Raspberry Pi. The online version can save projects to the Scratch server if you log in.

06 Exploring the Scratch community

You will probably want to find out lots more about how to use Scratch, and there are lots of tutorials if you select the Tutorials section in the top menu bar of the desktop or online version. You can also get lots more help from the main Scratch web page at **scratch.mit.edu/ideas**. If you look at the Explore section on the website, you will be able to find lots of projects that other people have made and if you find one you like, you can see how they did it by selecting the 'See inside' button at the top-right corner.

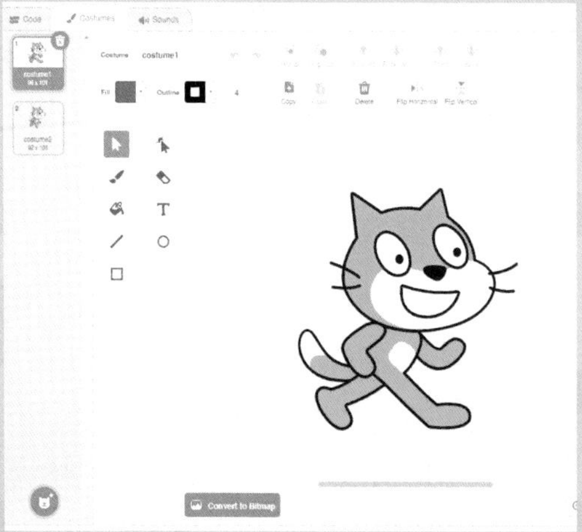

◀ The costume editor allows you to edit and design graphics for your games

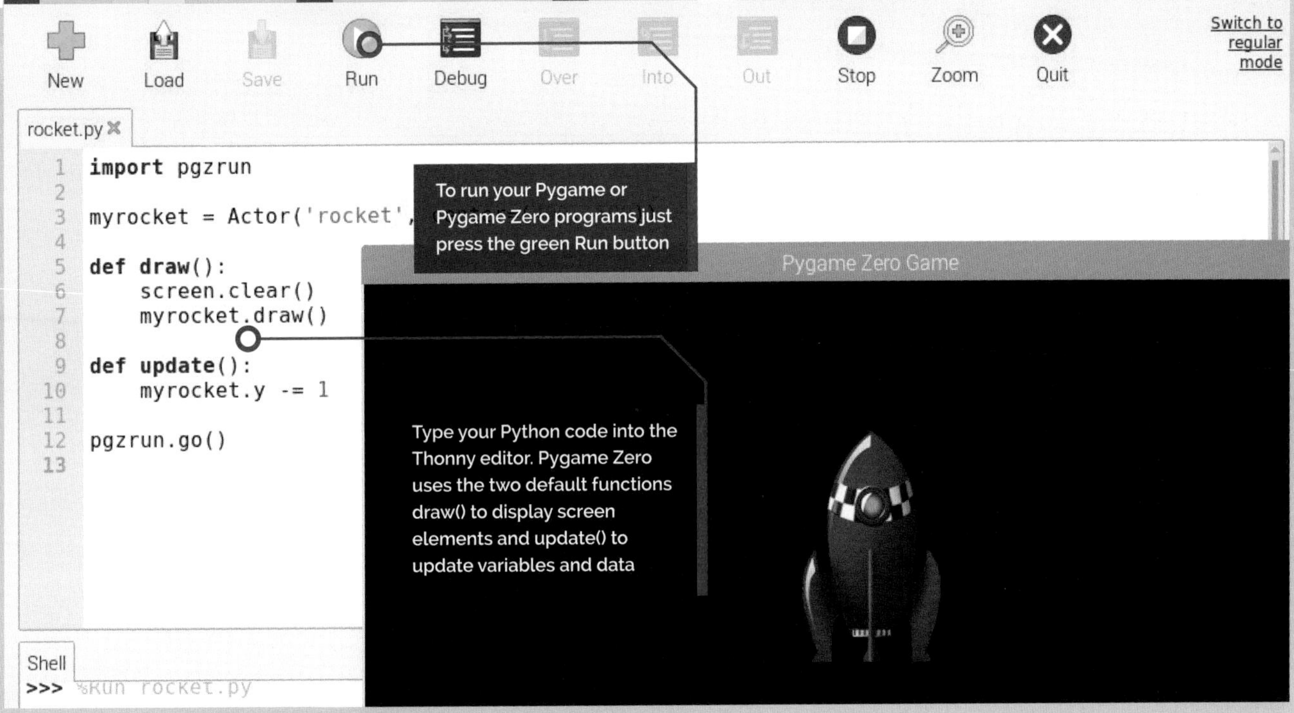
New
Load
Save
Run
Debug
Over
Into
Out
Stop
Zoom
Quit
Switch to regular mode
rocket.py
import pgzrun
myrocket = Actor('rocket',
def draw():
screen.clear()
myrocket.draw()
def update():
myrocket.y -= 1
pgzrun.go()
To run your Pygame or Pygame Zero programs just press the green Run button
Pygame Zero Game
Type your Python code into the Thonny editor. Pygame Zero uses the two default functions draw() to display screen elements and update() to update variables and data
Shell
>>> %Run rocket.py

CODE PYTHON GAMES WITH PYGAME

Raspberry Pi can be used to make some super games and Pygame gives you a great head start

> **Top Tip**
>
> **Mix and Match**
>
> Even if you start your program with Pygame Zero, if you need a function from Pygame you can include parts of the Pygame module too!

One of the best ways to get started with text-based programming on your Raspberry Pi is to jump straight into Pygame or Pygame Zero. These are both available with the Python programming language and all three are already installed by default with Raspberry Pi OS. If you are not familiar with Python, you can get it running from the Programming menu by selecting the Thonny Python IDE. This will open up an editor to use Python 3. Python is easy -to learn and read, and this article will show you how to use it with Pygame and Pygame Zero.

01 First Pygame Zero

Pygame Zero was designed to require as little code as possible to get a game running. If you launch the Thonny Editor (IDE) and type `import pgzrun` to load the Pygame Zero module and then after that, write `pgzrun.go()` to start the game, you can then save the file and run it (with the green play button). If you have typed the code correctly, you will see a black window appear titled 'Pygame Zero Game'. You have written your first Pygame Zero game! It's not a great game yet but that's all you need to get the game engine running.

02 More than zero

Let's get a graphic moving on the screen. You will need to find a suitable image to use, perhaps a spaceship or little green man. Have a look at the 'Graphics Resources' section near the end of this feature about where to find graphics. A PNG is best; you can find our rocket at **magpi.cc/rocketart**. Now make a subdirectory in the same place as you saved your Python file and call it **images** and put your graphic file inside that directory. Now load that graphic into an Actor object in your code. Name your graphic file **rocket.png** (you must keep to lower-case letters) and load it by typing `myrocket = Actor('rocket', center=(400, 500))`.

03 Seeing the rocket

Now to get our rocket to display on the screen, we need to add some code to draw it. We do this with a `draw()` function, so type `def draw():` and press **RETURN**, then type `myrocket.draw()`. Then, to make the rocket move up the screen, we need to add an `update()` function by typing `def update():` and underneath type `myrocket.y -=1`. If we save and run this program, we should see the rocket moving up the screen. If you don't, check the **rocket.py** code to see what you have done differently. It may have some drawing left behind, so add `screen.clear()` at the beginning of the `draw()` function. If all is well, you have the start of your Pygame Zero game.

ROCKET.PY

▪ Language: Python

```
import pgzrun

myrocket = Actor('rocket', center=(400, 500))

def draw():
    screen.clear()
    myrocket.draw()

def update():
    myrocket.y -= 1

pgzrun.go()
```

04 Moving on to Pygame

Pygame Zero makes it very quick and easy to get games working on your Raspberry Pi, but if you want more flexibility you may find that Pygame is what you require. You will need to write a bit more code, but you will be able to access some functions like using game controllers. To start a Pygame program, you will need to import the pygame module using `import pygame` and then after that, make a call to `pygame.init()`. This starts the game engine off, but we won't see anything happen if we run it.

05 Making a screen

We make a screen for our game by calling a function called `pygame.display.set_mode()` and give it the width and height that we want the screen to be. Once that is set up, we will need to start a loop (in this case a `while` loop) to check that the program is still running – and in the loop we blank the screen, draw our graphics on an invisible buffered screen, and then flip the screen from the buffer to the visible screen. All this keeps happening until the user exits the program by using the window close icon. Have a look at **pgtest.py** to see how all this is done with Pygame.

◀ Lots of Pygame game developers share their creations online like this game called Dynamite

PGTEST.PY

▪ Language: Python

```
import pygame
pygame.init()

screen = pygame.display.set_mode([400, 400])
running = True

while running:
    # Get events from the user
    for event in pygame.event.get():
        if event.type == pygame.QUIT:
            running = False
    # Fill the screen
    screen.fill((0, 0, 0))
    # Draw a red circle at 200,200 with
radius of 50
    pygame.draw.circle(screen, (255, 0, 0),
(200, 200), 50)
    # Switch from buffered screen to visible
    pygame.display.flip()

# Quit the program
pygame.quit()
```

06 Taking it further

Both of these examples are very simple, just to get you started, but there are lots of amazing games that can be made with Pygame and Pygame Zero. Over the years, *The MagPi* magazine has featured many tutorials about making games in Python and has even produced three books dedicated to teaching Python Games by example: *Retro Gaming with Raspberry Pi* (**magpi.cc/retrogaming**) and *Code the Classics – Volume 1* (**magpi.cc/ctc1**) which have many Pygame Zero example games, and the other is called *Essentials – Make Games with Python* (**magpi.cc/essentialgames**) which takes you through game creation with Pygame.

Top Tip

Watch your naming

When you save your code, don't call it 'pygame' or Python will think that you are referring to the pygame module.

WHERE TO GET IDEAS

Did you know that *The MagPi's* sister magazine, Wireframe, features a section called Source Code every month with Pygame Zero game examples? **wfmag.cc**

MAKE GREAT GAMES WITH APPGAMEKIT

With AppGameKit you can develop professional-looking games, not just for Raspberry Pi but also for desktop and even mobile devices

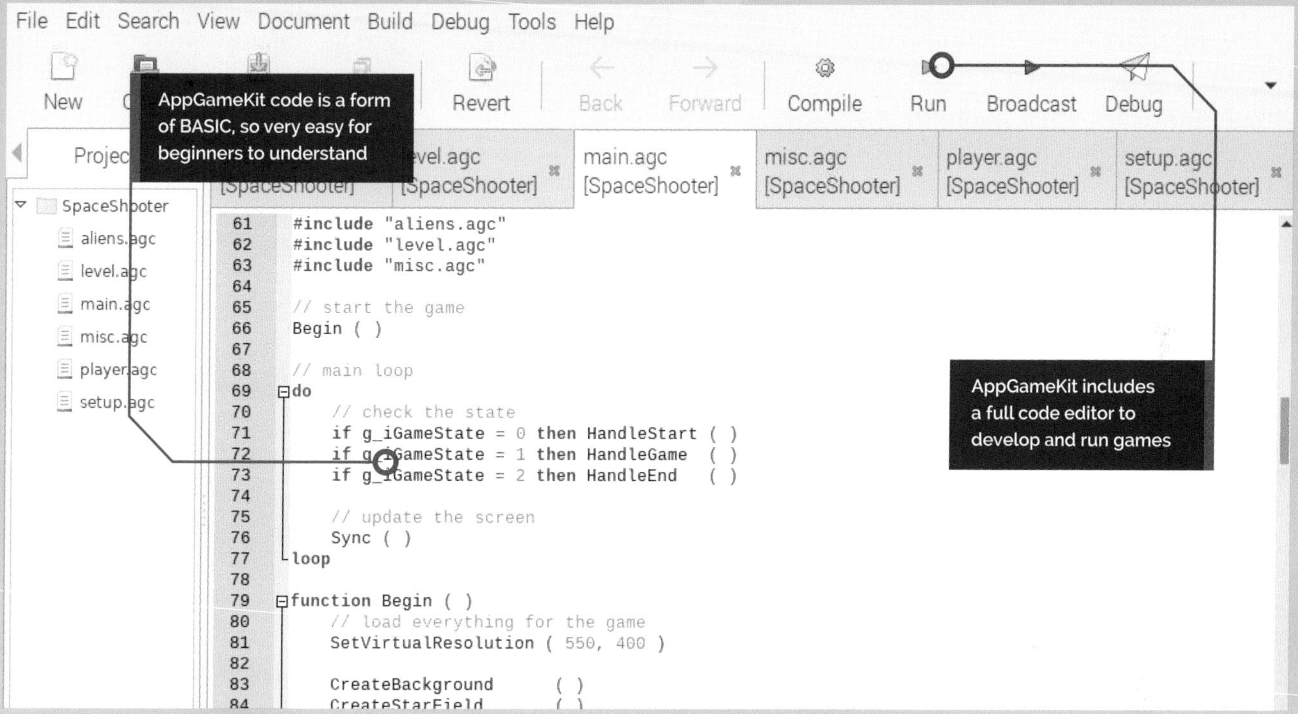
File Edit Search View Document Build Debug Tools Help
New
Revert
Back
Forward
Compile
Run
Broadcast
Debug
AppGameKit code is a form of BASIC, so very easy for beginners to understand
main.agc [SpaceShooter]
misc.agc [SpaceShooter]
player.agc [SpaceShooter]
setup.agc [SpaceShooter]
SpaceShooter
aliens.agc
level.agc
main.agc
misc.agc
player.agc
setup.agc
#include "aliens.agc"
#include "level.agc"
#include "misc.agc"
// start the game
Begin ( )
// main loop
do
// check the state
if g_iGameState = 0 then HandleStart ( )
if g_iGameState = 1 then HandleGame ( )
if g_iGameState = 2 then HandleEnd ( )
// update the screen
Sync ( )
loop
function Begin ( )
// load everything for the game
SetVirtualResolution (550, 400)
CreateBackground ( )
AppGameKit includes a full code editor to develop and run games

Top Tip

GPIO pins

If you are feeling adventurous, you could try out the AGK features that allow you to read and write to and from the GPIO pins.

AppGameKit provides a cross-platform development system that was originally for PC desktops, but recently it has become available to download free for Raspberry Pi. You can use the same system on other platforms too, and develop on one system to run on a different one. You can even publish your games and earn money without paying any royalties. The engine has many tools to help you build your game, like 2D sprites, 3D, physics, sound, and even virtual reality. This guide will get you started with AppGameKit so that you can explore all the features.

Compatibility alert

Until recently, AppGameKit was compatible with all Raspberry Pi computers, but at the time of writing, it is difficult to get running on Raspberry Pi 4. Some system updates are needed for other Raspberry Pi computers, even with the latest version of Raspberry Pi OS. Make sure you have backed up any data from your Raspberry Pi microSD card before you start.

01 Get the download

First, we need to get the AppGameKit files. You'll need to go to the website **appgamekit.com** and sign up for an account. When that's done, go to the 'AppGameKit For Raspberry Pi' section in the 'Classic' menu item and download the editor files (they are free). Double-click the gzip file to open it and extract the files to somewhere suitable like your home directory. When it's unpacked, you will see a directory called **AGKPi**. Inside that, you will find the AGK launcher. Double-click to open the editor. If you want to see an error log when it's running, select 'Execute in Terminal' when prompted.

02 Doing the update

If you try to run any of the samples provided with AGK, you may find that you get some errors. This may only be an issue at the time of writing as there are regular updates available.

From a Terminal window, enter `sudo apt update`, followed by `sudo apt upgrade`, just to make sure we have everything up to date. Then, if you are getting errors about libgles2 (graphics library), type `sudo apt install libgles2-mesa libsdl2-dev`, which will install the necessary libraries. Then enter `sudo rpi-update` – this is a firmware update, so a bit more extreme than the usual updates, and the reason why you should make a backup of your memory card before issuing this command. Now reboot your Raspberry Pi.

03 Load a sample

A good sample to start with, to make sure everything is working, is the Space Shooter game. Select the Open icon on the toolbar of the editor and then navigate to the **SpaceShooter** directory, which is found in the **Games** folder inside **Projects**. Open the .agk file and you will see several files open in the editor. AGK uses a language very much like BASIC, so if you have used BASIC before you should be right at home. If you haven't learnt BASIC, then it's quite easy as it was designed for beginners.

04 Run the game

If everything has gone well with the install and updates, when you press the green Run arrow you should see a window open up titled AGK and a Start Game screen with spaceships floating about. If you don't see that, then check the Terminal window that launched the editor to see if there are any errors. You may see some warnings there anyway as some of the shader modes are not supported on Raspberry Pi, but the game should work fine. Start the game by clicking the screen and move the player ship up and down with cursor keys.

05 Make your program

Now you have the editor building a game, why not start your own? Start a new project by clicking the New icon on the toolbar. You will be asked for a name for the new project and a base path. Select the folder icon to the right of the base path input box, and navigate to somewhere suitable inside your home directory. Select Create and you will see a new file called **main.agc** open in the editor. In that file, there will already be a listing of the base code you need to start your game. Run it and you will see a black window open with the title 'test'.

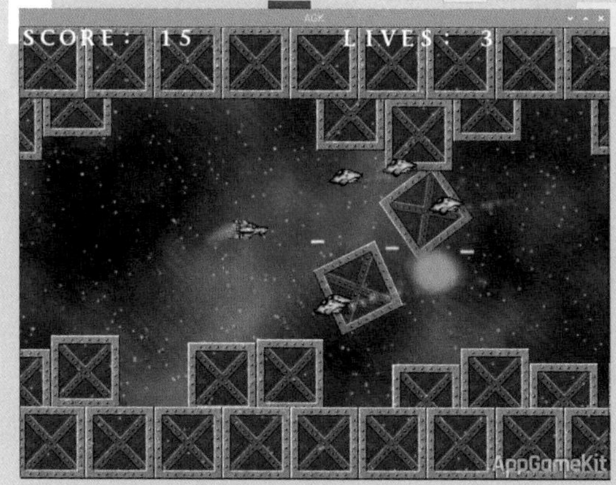

With the AppGameKit samples, you can quickly see how to build many types of games

06 The sky's the limit

Have a look through the samples to see the range of what AGK can do. You will find a huge range of tutorials at **magpi.cc/agkyt** and there is a full user guide at **magpi.cc/agkguide**. There is also an active and helpful community forum at **magpi.cc/agkforum** where you will find more hints and tips to help you on your way. If you are having a problem with something, you'll find someone who has solved it and will tell you how. Don't be afraid to get stuck in and just start coding: the compiler will give you feedback on anything you get wrong.

Top Tip

Raspberry Pi 4

Feeling brave and want to run AGK on a Raspberry Pi 4? Check out the forum post at: **magpi.cc/agkpi4** for instructions.

You'll Need

- AppGameKit: **appgamekit.com**
- Raspberry Pi SDK: **appgamekit.com/agk-pi**

GRAPHICS RESOURCES

There are many graphics resources that are free to download. You can get images, animations and programs. Here are a few places to visit:

- **opengameart.org** has a wide range of artwork for backgrounds and character images to include in your games free of charge.
- **spriters-resource.com** specialises in sprites, which are the characters to include in games. They are often available in sprite sheets which have all the frames in one image file.
- **free3d.com** has many free (and paid-for) 3D models for you to download and use. There are models for just about any situation, some of them specifically designed for games.
- **gimp.org** is a great image manipulation program and should be all you need for creating 2D graphics for your games. It can be installed using `sudo apt install gimp` in a Terminal window.
- **blender.org** is best for creating a game with 3D graphics. Install Blender free from your Terminal with `sudo apt install blender`. Discover a range of Blender projects on Raspberry Pi's website (**magpi.cc/blenderprojects**).

MAKE ADVENTURE GAMES WITH REN'PY

This game engine is for storytelling. Use Raspberry Pi to combine words, images, and sounds to create interactive visual novels and life simulation games

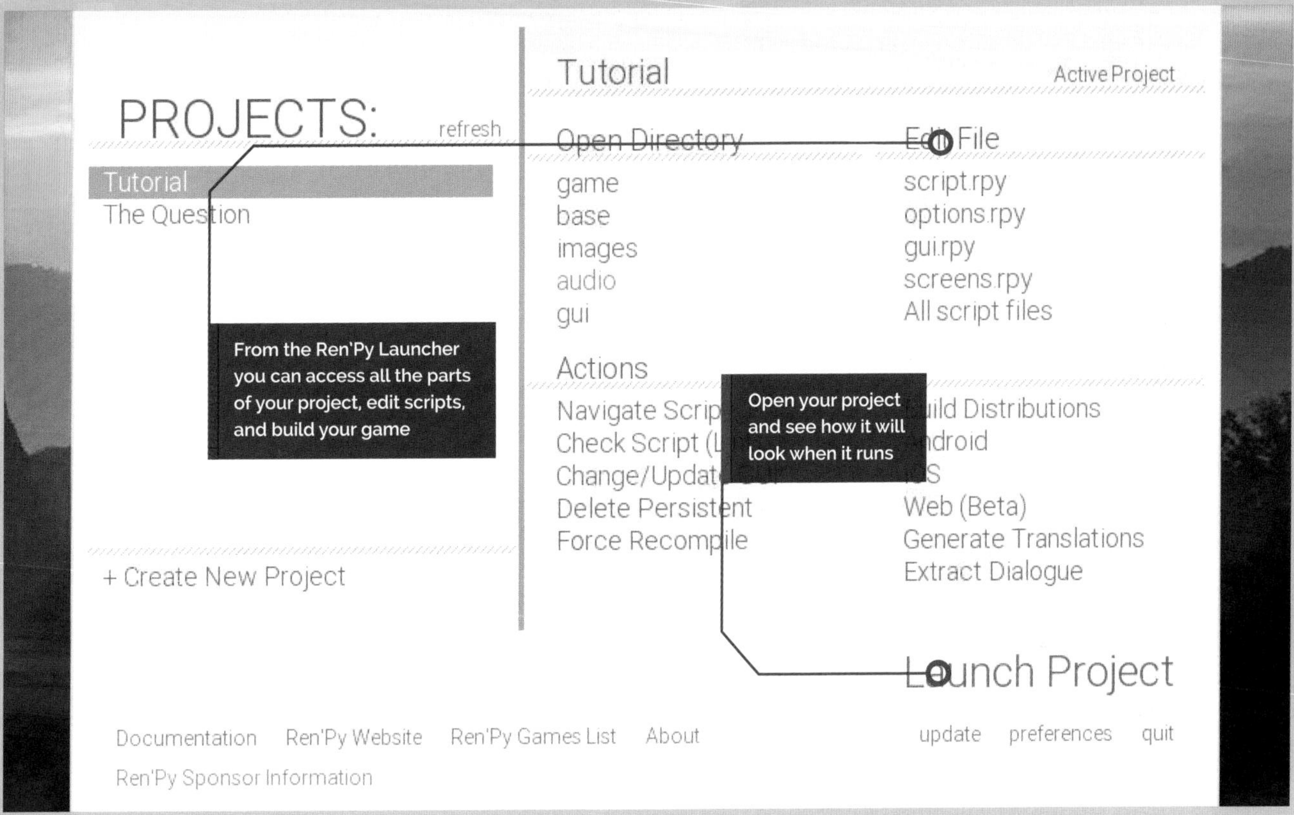
PROJECTS:
refresh
Tutorial
The Question
+ Create New Project
Tutorial
Active Project
Open Directory
game
base
images
audio
gui
Edit File
script.rpy
options.rpy
gui.rpy
screens.rpy
All script files
Actions
Navigate Script
Check Script (L
Change/Update
Delete Persistent
Force Recompile
Build Distributions
Android
iOS
Web (Beta)
Generate Translations
Extract Dialogue
Launch Project
Documentation
Ren'Py Website
Ren'Py Games List
About
Ren'Py Sponsor Information
update
preferences
quit

From the Ren'Py Launcher you can access all the parts of your project, edit scripts, and build your game

Open your project and see how it will look when it runs

Top Tip

Embed Python

Ren'Py scripting is quite similar to Python, but if you need to embed a Python program inside your Ren'Py game, you can do that too.

Ren'Py is open-source and free to download and use. You can even share your creations without paying a penny in royalties or licences. Ren'Py includes a simple scripting language to control the flow of your story and add interactivity to the pages. The engine also includes a wide selection of animation and transition effects to bring your games and graphic novels to life without needing to learn complicated animation software and supports the most common graphics and sound formats like JPG, PNG, MP3, and a whole lot more.

01 Get the files

First, download the install files from the Ren'Py website at **renpy.org/latest.html**. You will need the .bz2 version for Raspberry Pi. When it has downloaded, double-click to open the archive and extract it to a suitable place such as your home directory. You will also need to download and extract Raspberry Pi support files from the Additional Downloads section. Once this is all in place, you will find a file in the directory you have extracted called **renpy.sh**. Double-click this file and select 'Execute'. After a few seconds, you will see the Ren'Py Launcher open.

02 Tutorial time

Ren'Py includes a getting started tutorial, which is probably the best place to begin. By selecting the Tutorial project from the launcher, you will be introduced to Ren'Py's features by Eileen. She will show you how to start a new project and the ways to set colours and screen sizes. There are also sections in the tutorial to

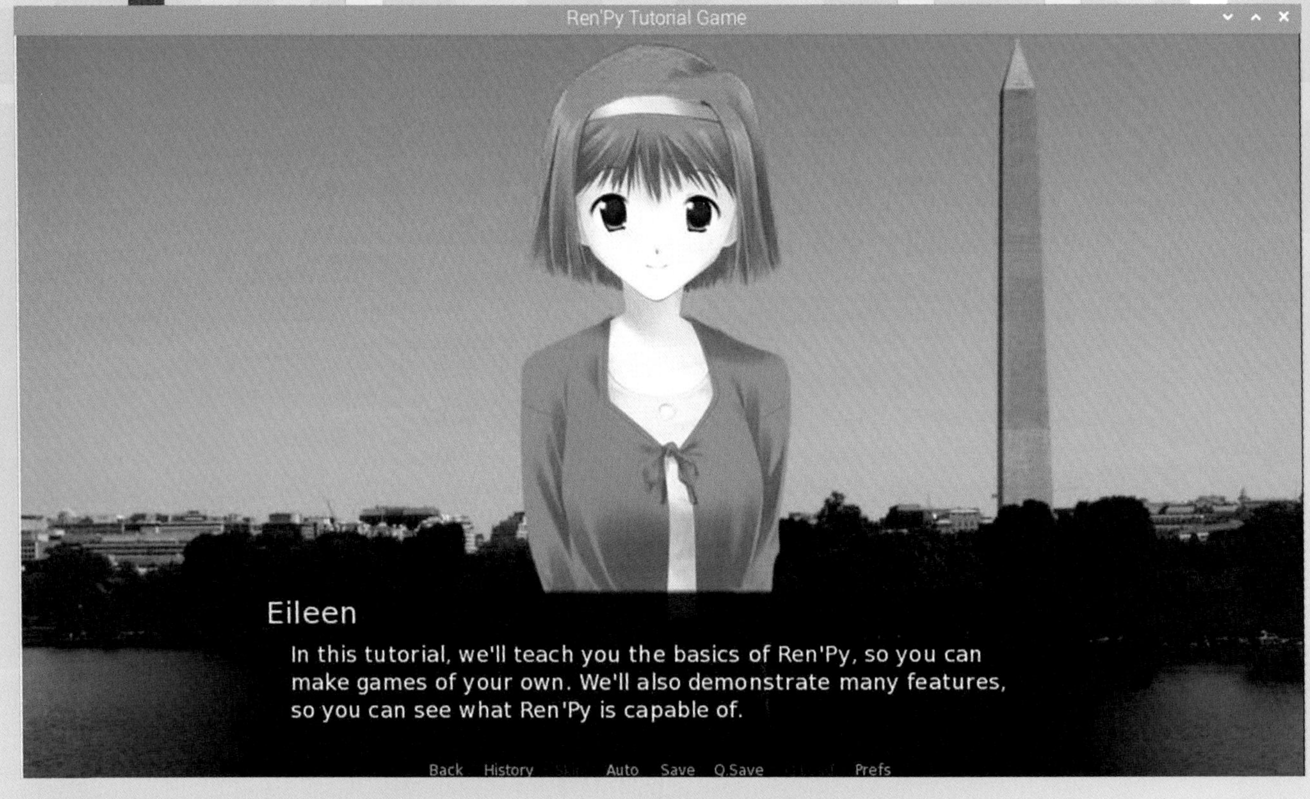
Ren'Py Tutorial Game
Eileen
In this tutorial, we'll teach you the basics of Ren'Py, so you can make games of your own. We'll also demonstrate many features, so you can see what Ren'Py is capable of.
Back History Auto Save Q.Save Prefs

▲ Ren'Py includes a tutorial where Eileen talks you through all the features of the system

cover adding your images, text, and sound to your pages. It then goes on to creating interactions and transitions to make your game engaging for your audience. Have a look at the Choices and Python section to see how scripting is used to ask questions and branch to different options.

03 Let's make a game

Going back to the Ren'Py Launcher, start a new project with the 'Create New Project' link on the left-hand side. You will then be asked where you want to save your project and what it should be named. Next, choose what screen resolution you want your production to use and the colour scheme that you would like. After a short pause for processing, your project will be created and listed with the tutorial in the Projects section in the Launcher.

04 Let's get scripting

Start scripting the game by selecting the 'script.rpy' option under Edit File in the Launcher. It will ask you to select the editor you want to use and then open the script. From there you can make changes to the default script. When you want to test your changes, select your project and Launch Project, then select 'Start' from the list on the left of the window that opens. If you make changes to your script, you can then press **SHIFT+R** to reload your script and start the game again. If you need further help, select the Documentation link at the bottom left of the Launcher window, or check out the forums at **magpi.cc/renpyforum**.

SOUND RESOURCES

If you need to find sounds for your games, you can get a whole range of sound effects and background soundtracks from **freesound.org**, **zapsplat.com,** or **musopen.org/music**, all of which provide free downloads.

You may need to edit your sounds, in which case use Audacity – available for you to install using `sudo apt install audacity` from your Terminal window.

▼ Audacity enables you to edit sound files in a variety of formats such as WAV and MP3

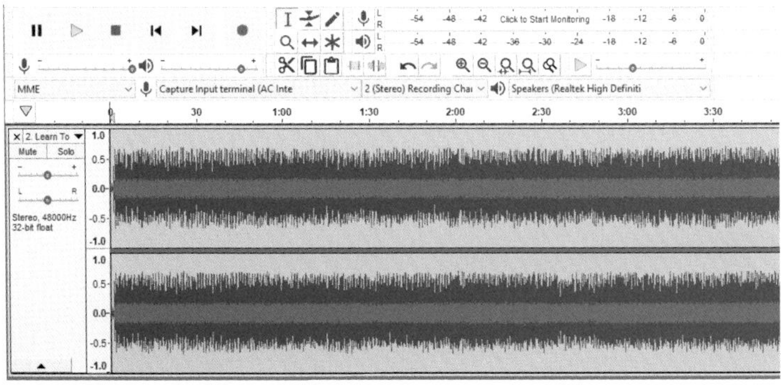

UL LISTED

Industrial Raspberry Pi

ComfilePi

The ComfilePi is a touch panel PC designed with high-tolerant components and no moving parts for industrial applications. It features a water-resistant front panel, touchscreen, color LCD (available in various sizes), RS-232, RS-485, Ethernet, USB, I2C, SPI, digital IO, battery-backed RTC (real-time clock), and piezo buzzer.

Use the rear-panel 40-pin GPIO header to expand its features and capabilities with additional I/O boards. The ComfilePi is UL Listed and employs Raspberry Pi Compute Module.

POWERED BY Raspberry Pi

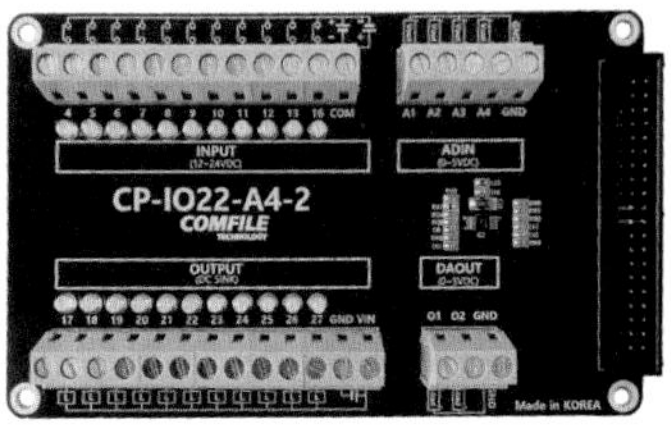

COMFILE TECHNOLOGY

isit www.comfiletech.com © copyright COMFILE Technology, Inc. ALL RIGHTS RESERVED

Pico W

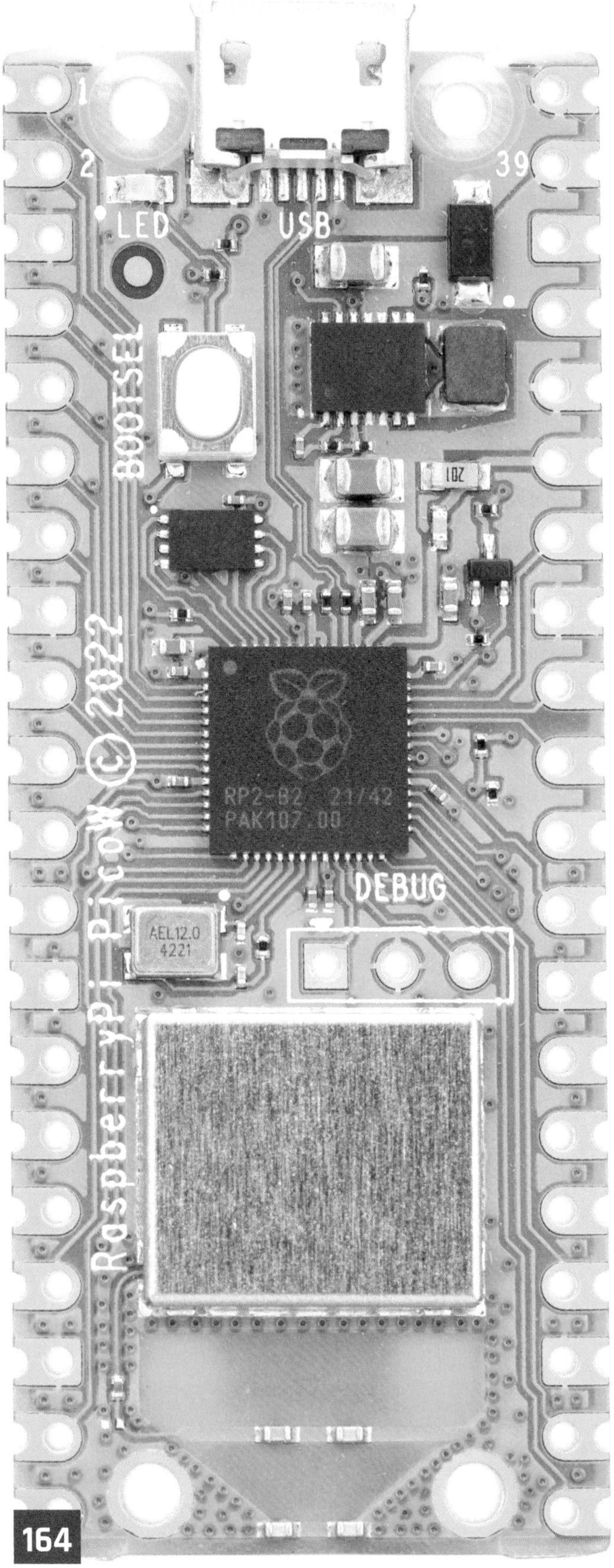

164

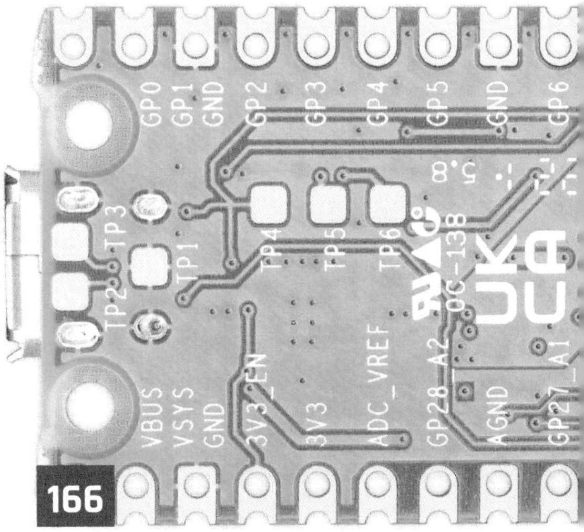

166

162
BOOTSEL
Raspberry Pi Pico W © 2022
GND
GP10
GP11
GP12
GP13
GND
GP14
GP15
FCC ID:2ABCB-PICOW
IC:20953-PICOW
Uses technology
licensed from
Proant AB
GND
GP21
GP20
GP19
GP18
GND
GP17
GP16

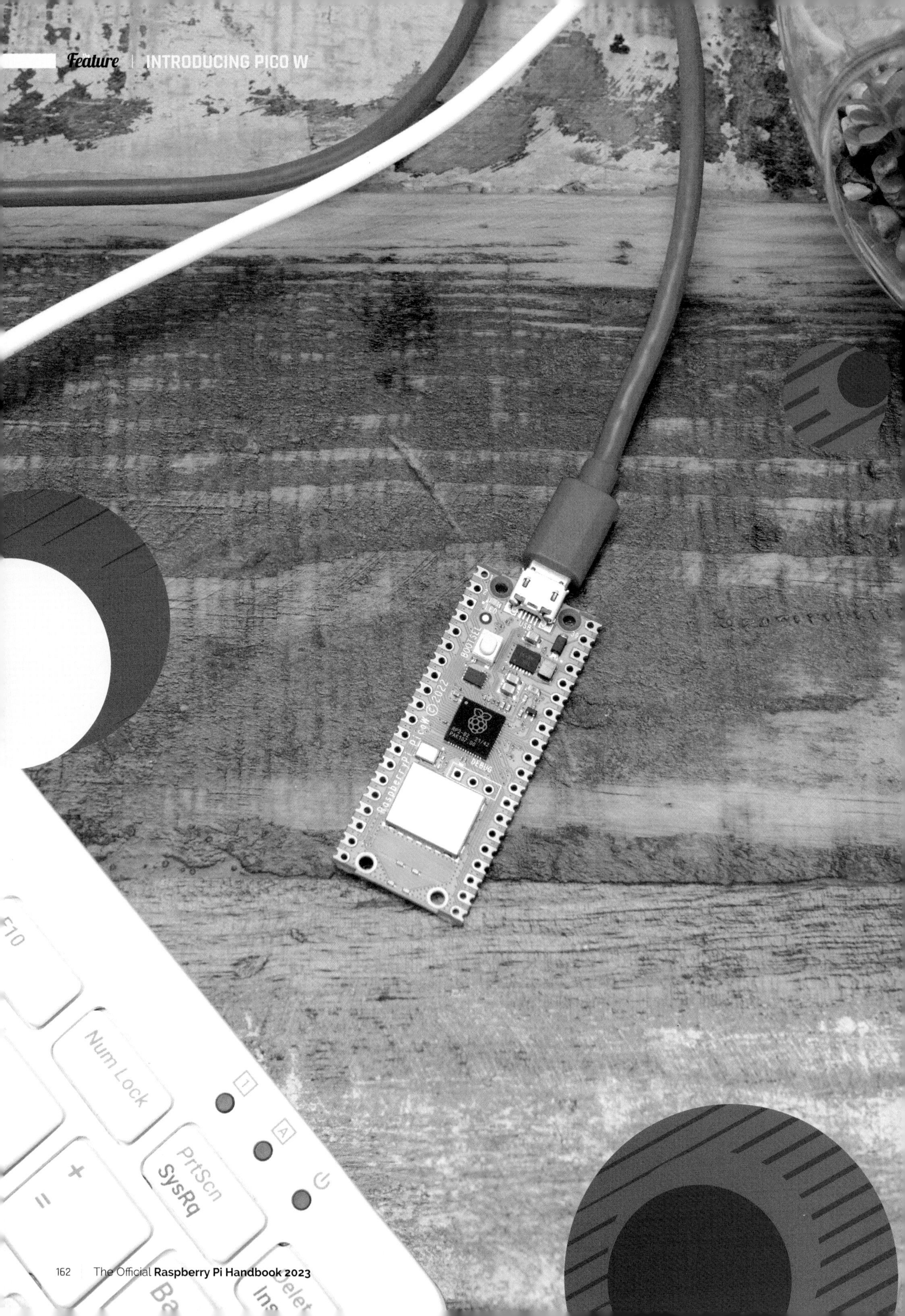
BOOTSEL
Raspberry Pi Pico W © 2022
DEBUG
F10
Num Lock
PrtScn
SysRq

INTRODUCING PICO W

Raspberry Pi's RP-2040 microcontroller development platform gets native wireless connectivity. By **Lucy Hattersley**

Raspberry Pi Pico has been a standout star of recent years. The tiny Pico development board contains a microcontroller chip designed by Raspberry Pi called 'RP2040'.

Programs are loaded onto Pico via a separate computer, such as a larger Raspberry Pi board, and run automatically as soon as the power is turned on. The GPIO pins provide input and output and a vibrant community and maker scene have built up to get the most from Pico.

We are incredibly proud to show you the all-new Raspberry Pi Pico W. As the 'W' moniker suggests; Pico W is packing wireless connectivity.

This tiny board now houses a silver square module containing an Infineon CYW43439 (**magpi.cc/CYW43439**) device. This enables RP2040 to connect to the internet via a wireless interface using an 802.11n, single-band (2.4GHz) connection. The module sits alongside an familiar trapezoidal antenna on the edge of the board. All of this required a clever engineering design, and we're going to reveal all in this feature.

Raspberry Pi Pico remains a low-cost development platform. And you can pick up Pico W from just $6 (around £5). It is programmed by connecting to another computer, such as a Raspberry Pi 4 or 400, or a similar Linux, macOS, or Windows computer. Programs are created using the C/C++ SDK or MicroPython.

> "All of this required a clever engineering design"

Files are drag and drop placed onto Pico W via the USB connection or SWD pins, and you can interactively debug code running on Pico W.

There's never been a better time to get into programming with Raspberry Pi Pico W. Pico can be used to control many hardware projects of your own design, and there is a huge ecosystem of kits and components.

We believe that wireless LAN is a game-changer for Pico. It will enable the device to connect from the edge to your network and the wider internet, and pass data back for storage and analysis.

We can't wait to see what folks make with Pico W.

GET TO KNOW PICO W

4 BOOTSEL
This button is used to connect Pico W to a computer. It then appears as a drive on the other machine.

3 SPI EEPROM
This small EEPROM chip houses the code used by Pico W to perform its operations. The programs you upload to Raspberry Pi Pico are stored here.

2 OSCILLATOR
The AEL1210CS oscillator provides the 12MGz clock frequency that keeps RP2040 ticking along.

1 WIRELESS LAN
The Infineon CYW43439 wireless chip is housed within this silver-coloured package and provides a single band of 2.4GH wireless LAN (802.11n).

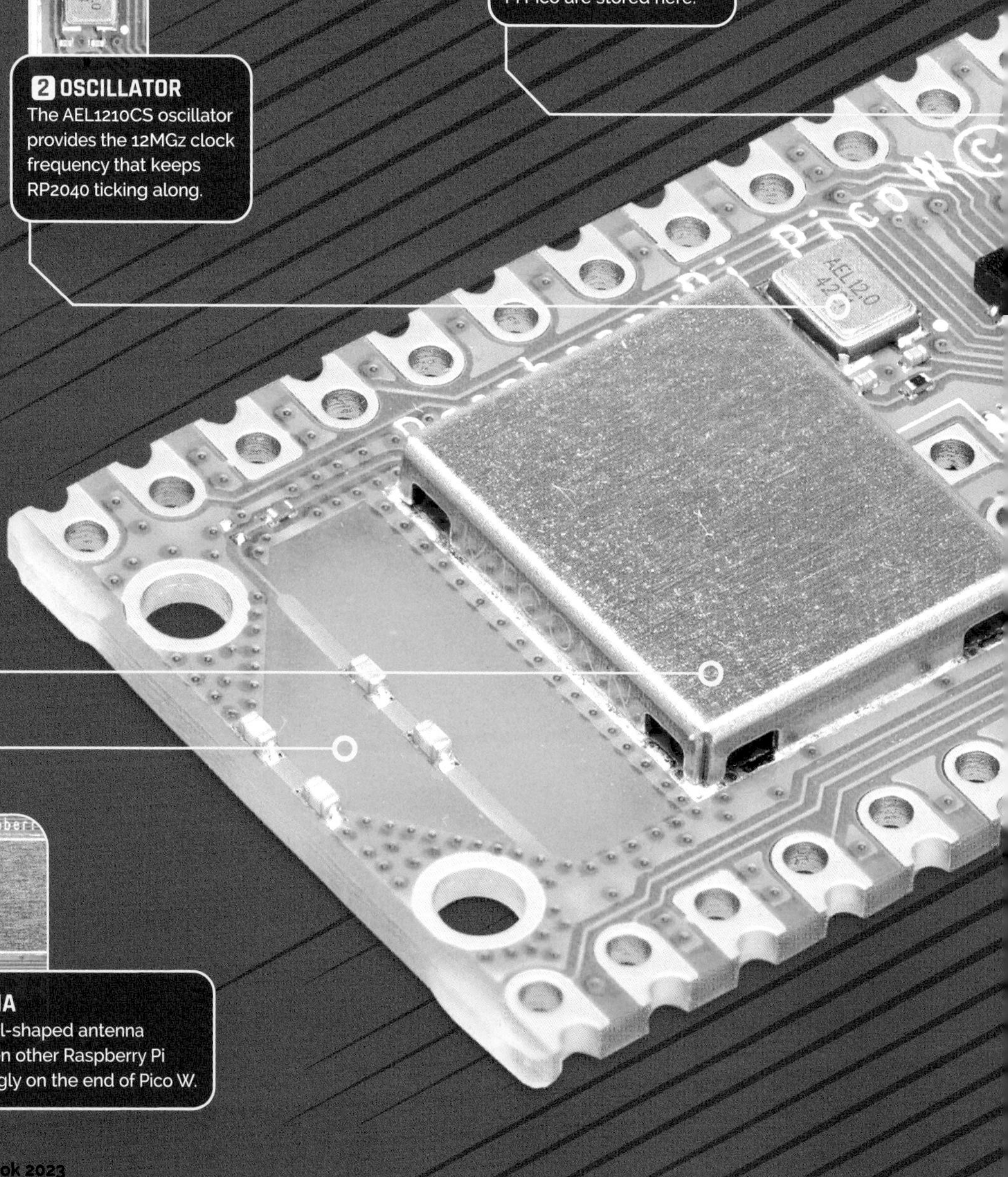

10 ANTENNA
The trapezoidal-shaped antenna design found on other Raspberry Pi boards fits snugly on the end of Pico W.

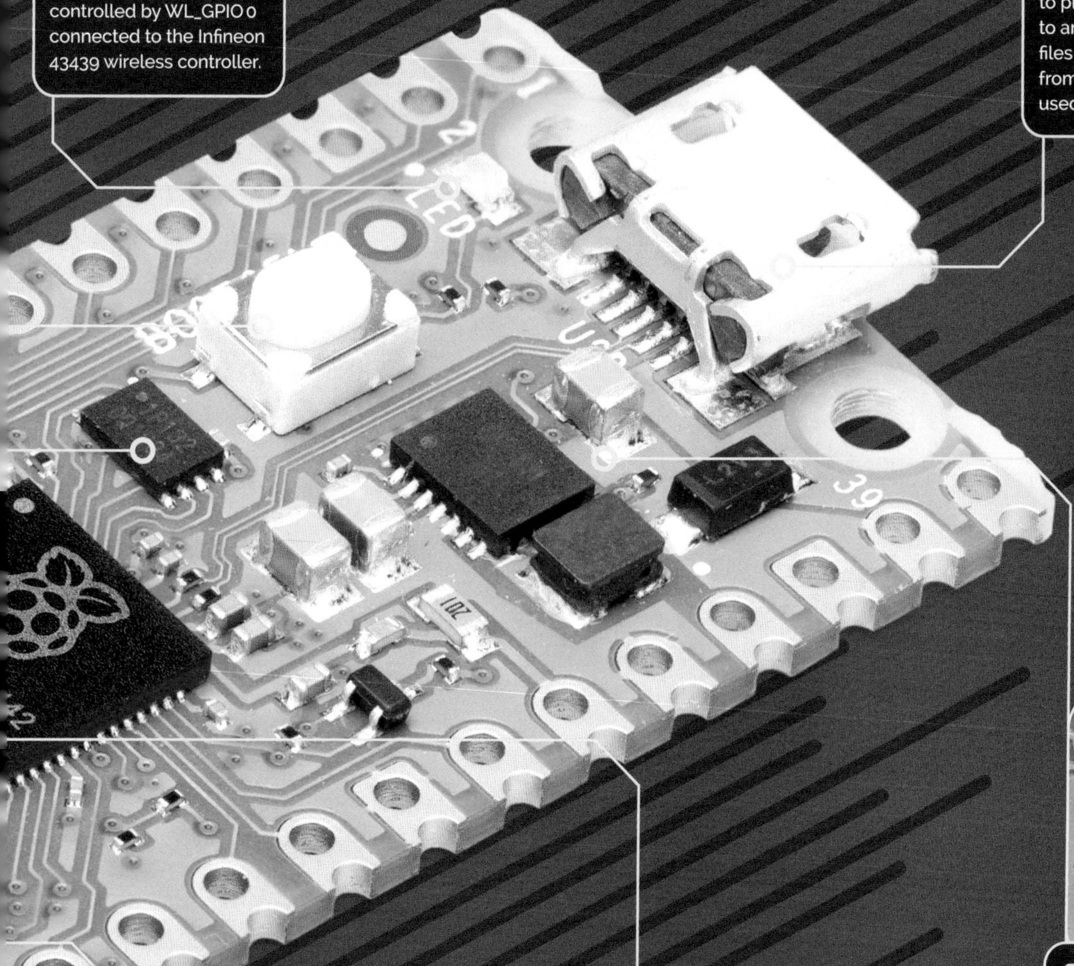

5 LED

A single LED remains on the board and is now controlled by WL_GPIO 0 connected to the Infineon 43439 wireless controller.

6 MICRO USB

A single micro USB port is used to provide power and connect to another computer so code files can be transferred to and from Pico W. This can also be used for direct code debugging.

7 POWER MANAGEMENT

These chips form the power management for Raspberry Pi Pico, which is fixed at 3.3V.

8 RP2040

The beating heart of Pico is the RP2040 microcontroller, appropriately housed in the black package and adorned with a Raspberry Pi logo.

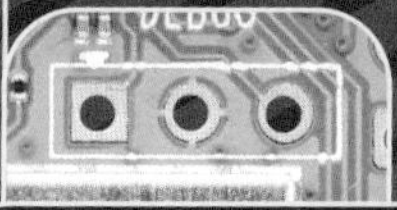

9 DEBUG PINS

The three debug pins have been relocated from the edge of the board to a more central location.

TAKE A LOOK AT THE BACK OF PICO W

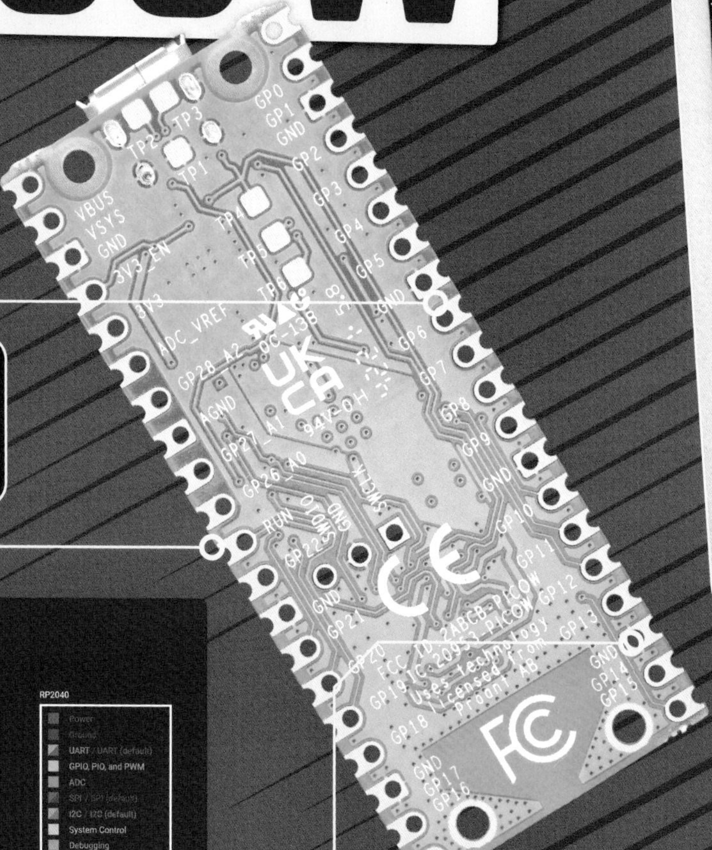

1 GPIO PINS
On the two sides of Raspberry Pi Pico W are the 40 GPIO pins used to provide I/O between Pico W and other hardware. All of the GPIO pins are labelled on the rear of Pico W.

2 CASTELLATIONS
The GPIO pins are on a 21 mm × 51 mm 'DIP' style 1 mm thick PCB, with 0.1˝ through-hole pins, also with edge castellations.

3 SQUARE GROUND PIN
Look closely at the ground pins. These have square edges instead of round ones, making them easier to locate when you are prototyping.

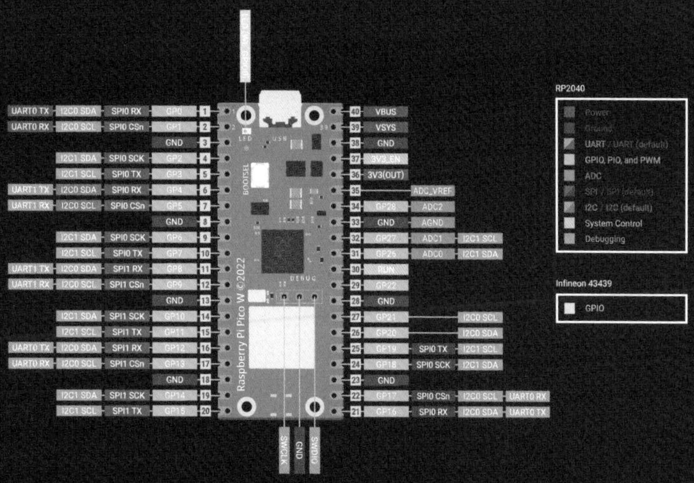

RASPBERRY PI PICO W GPIO GUIDE

RASPBERRY PI PICO W SPECIFICATIONS

- RP2040 microcontroller with 2MB of flash storage
- On-board single-band 2.4GHz wireless interfaces (802.11n)
  - WiFi 4 (802.11n), single-band (2.4 GHz)
  - 20MHz channels
- Micro USB B port for power and data (and for reprogramming the flash)
- 40-pin 21 mm × 51 mm ‘DIP’ style 1 mm thick PCB, with 0.1-inch through-hole pins also with edge castellations
- 3-pin ARM Serial Wire Debug (SWD) port

For more information see the Raspberry Pi Pico W datasheet (**magpi.cc/docs**)

RP2040 MICROCONTROLLER SPECIFICATIONS

- Dual-core Cortex M0+ at up to 133MHz
- 264kByte multi-bank
- High-performance SRAM
- External quad-SPI flash with eXecute In Place (XIP) and 16kByte on-chip cache high-performance full-crossbar bus fabric
- On-board USB 1.1 (device or host)
- 30 × multi-function General-Purpose IO (4 can be used for ADC)
- 12-bit 500 ksps analogue-to-digital Converter (ADC)
- Various digital peripherals
- 2 × Programmable IO (PIO) blocks, 8 × state machines total

For full details of the RP2040 microcontroller, please see the RP2040 datasheet (**magpi.cc/2040datasheet**)

PICO WH

A Pico W with a built-in header, called Pico WH, is also announced and will be available soon (this is an image of Pico H). In the meantime, it is possible to solder headers to Raspberry Pi Pico W. Header kits are available for just £1/$1 (**magpi.cc/picoheaders**). Headers make it easy to connect Pico to HAT-like hardware and breadboards for circuit prototyping.

USING RASPBERRY PI PICO W

Set up your Pico W and get onto the internet with MicroPython

Wireless support for Raspberry Pi Pico W is available using MicroPython, a coding language for microcontrollers based on Python. There will be support for C/C++ as well.

In this tutorial, we are going to connect Pico W to a Raspberry Pi using USB (you can use a non-Raspberry Pi computer with a different operating system if you prefer).

We will then install the latest version of MicroPython with wireless LAN support onto a new Pico W and use it to connect to a wireless network.

01 Update Pico W firmware

Pico W has a BOOTSEL mode that enables you to update the firmware via a USB port. Download the latest **firmware.uf2** file (**magpi.cc/picowuf2**) from the MicroPython documentation page.

Make sure your Raspberry Pi Pico W is not connected to a power source and hold down the white BOOTSEL button on the board. With the button held down, connect Pico W to your computer using a micro USB cable.

A drive called RPI-RP2 should appear in your computer's file system. Drag the **firmware.uf2** file to this drive. It will take a while for the file to copy across. Pico W will reboot when finished, and the RPI-RP2 drive will disappear and boot into MicroPython.

02 Connect over USB

When Raspberry Pi Pico W boots, it waits for you to tell it what to do. You can load a Python '.py' file from your computer, or interact directly with a read-evaluate-print loop (or REPL). MicroPython is equipped with a virtual USB serial port that can be accessed via the micro USB connector. Make sure that Pico W is connected to your computer via USB, and that you have not held down the BOOTSEL button during connection.

Your computer should notice Pico W's serial port as a character device, most likely:

```
/dev/ttyACM0
```

Open a Terminal window and use this command to list your serial ports:

```
ls /dev/tty*
```

There may be a lot of tty connections, but MicroPython's USB serial will start with **/dev/ttyACM** if you are using a Linux computer. If using macOS, it will have the extension '.usbmodem' with a number appended to the end.

Top Tip

Serial Wire Debug

Another way to communicate with Pico W is via the SWD (Serial Wire Debug) pins. See Chapter 5: Flash Programming with SWD in the Getting Started with Pico book (**magpi.cc/getstartedpico**).

03 Install Minicom

We're going to use Minicom to access the serial port:

```
sudo apt install minicom
```

And connect to Pico W using:

```
minicom -o -D /dev/ttyACM0
```

You'll Need

- Raspberry Pi Pico W **magpi.cc/picow**
- firmware.uf2 **magpi.cc/picowuf2**
- Raspberry Pi 4/400 (or alternative computer) **magpi.cc/raspberrypi400**
- Wireless LAN network
- Thonny IDE **magpi.cc/thonny**

The `-D` option and `/dev/` address are pointing Minicom to MicroPython's USB serial port, and the `-o` flag is a default 'just do it' option that stops Minicom initialising modem and lock files.

Top Tip

Minicom on Mac

If you are using an Apple Mac with a recent version of macOS, the serial will show up with a .usbmodem extension (followed by a number). First, install Homebrew (**magpi.cc/homebrew**) to install packages in macOS. Then enter:

```
brew install minicom
```

And connect to the board as below (replacing the number '1101' at the end with your number as listed in Step 2):

```
minicom -b 115200 -o -D /dev/tty.
usbmodem1101
```

04 MicroPython prompt

Press the **ENTER** key a few times and you should see a prompt:

```
>>>
```

This is the MicroPython prompt and you can enter commands here directly to your Raspberry Pi Pico W. If you press **CTRL+D**, you will reboot your Pico W and see something like:

```
MPY: soft reboot
MicroPython v1.18-454-g02496c997-dirty on
2022-05-18; Raspberry Pi Pico W with RP2040
Type "help()" for more information.
```

This is a good way to check the connection is working. The '>>>' prompt will reappear.

05 Connect to the network

We're going to connect Pico W to a local network using the network library. Enter the code from **network.py**, replacing the 'Wireless Network' and 'The Password' items with your own network and wireless LAN passcode.

network.py

> Language: **MicroPython**

DOWNLOAD THE FULL CODE: magpi.cc/internetpicow

```
import network
import time

wlan = network.WLAN(network.STA_IF)
wlan.active(True)
wlan.connect('Wireless Network', 'The Password')

while not wlan.isconnected() and wlan.status() >= 0:
    print("Waiting to connect:")
    time.sleep(1)

print(wlan.ifconfig())
```

disconnect.py

> Language: **MicroPython**

```
# Connect to another wifi
wlan.disconnect();
wlan.connect('Other Network', 'The Other Password')
```

> "We're going to connect Pico W to a local network using the network library"

Enter each line one at a time, pressing **ENTER** at the end of each one. When the code is entered, you should see the IP address your Raspberry Pi Pico W is using. For instance:

```
('10.3.15.196', '255.255.255.0',
'10.3.15.1', '10.3.15.1')
```

You can use the code from **disconnect.py** to disconnect and connect to a different network.

It is possible to connect to HTTP (Hypertext Transfer Protocol) websites using sockets or urequests. There is an example of sockets in the documentation. You can handle redirects and there is limited JSON support. And you can even use sockets to build a simple web server and control hardware using the web server.

Example code for using Raspberry Pi Pico W with the internet is being created and shared by Raspberry Pi.

Take a look at the Pico W documentation on Raspberry Pi's website (**magpi.cc/docs**).

Pico W resources

Make sure you download and bookmark these resource files

- Pico W datasheet **magpi.cc/picowdatasheet**
- RP2040 datasheet **magpi.cc/rp2040datasheet**
- Hardware design with RP2040 **magpi.cc/hdrp2040**

DOMINIC PLUNKETT AND LIAM FRASER ON RASPBERRY PI

PICO W

Senior principal hardware engineer Dominic Plunkett and software engineer Liam Fraser reveal all about Pico W

Dominic Plunkett

Liam Fraser

Adding wireless functionality to the ultra-small form-factor of Pico W was no easy task. "There was a big question about physical space," says Dominic Plunkett, senior principal hardware engineer at Raspberry Pi. "About whether we could get the wireless chip and the antenna on Pico."

A particular challenge was how to keep the GPIO pins and accommodate Infineon's CYW43439 technology. A problem that was avoided with some clever re-routing. "We could have potentially lost the bottom two GPIO pins on each side," he remarks. "We need four GPIO pins to control the wireless LAN."

▶ Thanks to the clever design, all 40 GPIO pins remain in the same position on the newly designed Pico W board

Finding a shape

Removing the GPIO pins around the antenna was tempting because it would free up space: "Antennas like space," explains Dominic while showing us the trapezoidal-shaped feature. "And getting rid of the bottom GPIO pins would have made it easier to connect the wireless chip," but it would have been a huge change for current users. "I didn't want to lose any of the peripheral GPIO pins to the end-user," says Dominic. People can add Pico W to an existing project without having to change anything and gain instant access to wireless technology.

"There was a lot of work to evaluate different antenna shapes and patterns to fit in the space, and to come up with the correct balance," says Dominic. This is one reason why the three debug pins moved from the edge of the board to a more central location: to make space for the antenna and prevent them from interfering with the wireless connection.

"After a bit of head-scratching and a long weekend of adding some secret extra resistors, we got away with using three pins to control the wireless LAN. So the three pins we've taken are the LED, power supply mode select, and the VBUS detect, and you get those back through the wireless LAN chip. It has some GPIOs which then go back to control the LED and power supply.

"It is a fine solution," affirms Dominic. "A lot of people don't use these pins. Maybe, on the first day, you might use the LED, but very quickly you go on to doing things with the I/O pins and that's why I felt it was important to try and make sure they have the existing I/O and that it was the same."

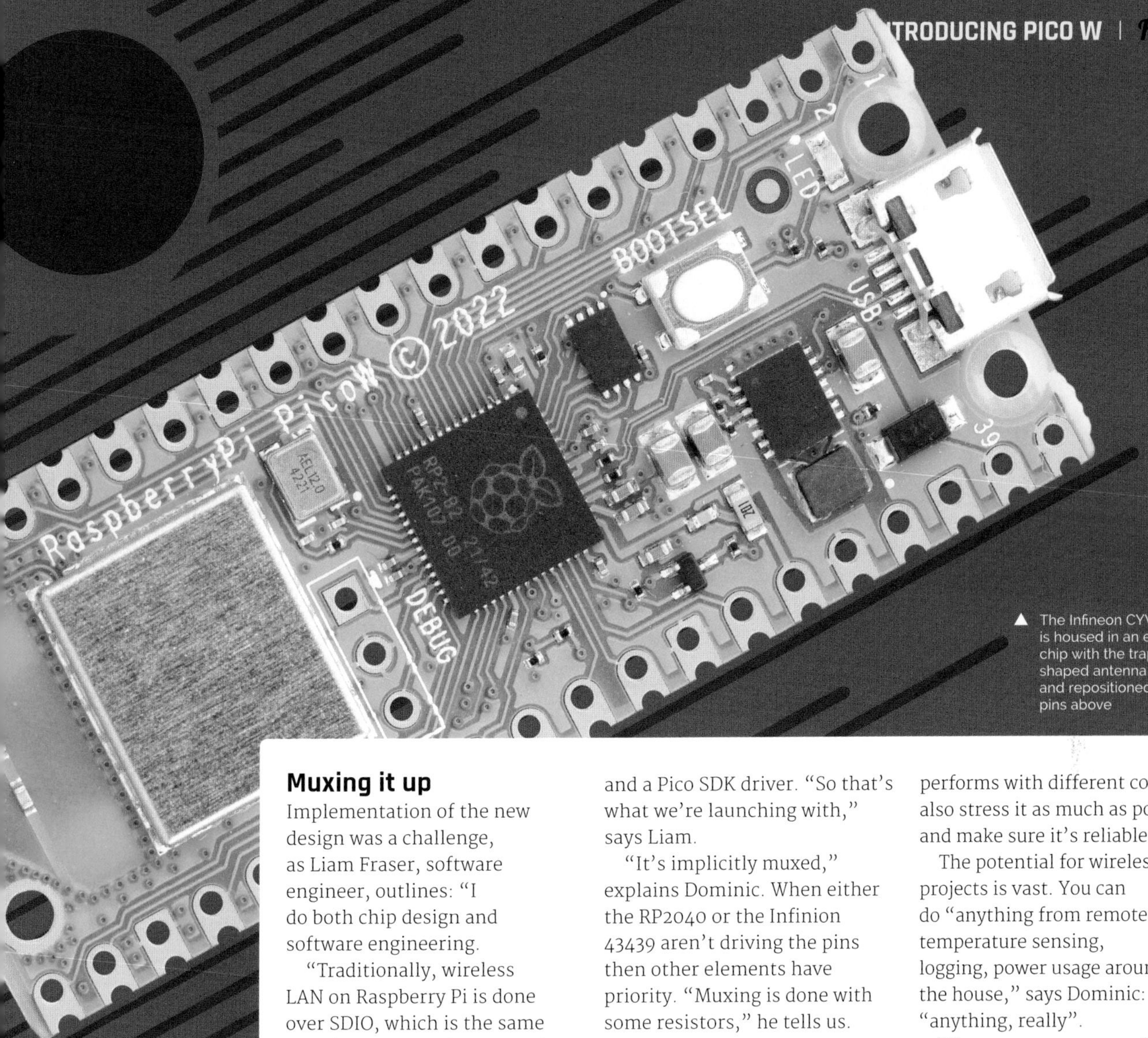

▲ The Infineon CYW43439 is housed in an enclosure chip with the trapezoid-shaped antenna below and repositioned DEBUG pins above

Muxing it up

Implementation of the new design was a challenge, as Liam Fraser, software engineer, outlines: "I do both chip design and software engineering.

"Traditionally, wireless LAN on Raspberry Pi is done over SDIO, which is the same interface as the microSD card. However, SPI lets you do it in fewer pins and then we've got muxing on the pins to make all of that fit into just the three pins.

"We are amongst the first to use the SPI mode," explains Liam. "One of the challenges was to come up with a code base that worked for MicroPython and Pico SDK," he tells us. "The Infineon supplied code required lots of libraries to be pulled in and only worked with FreeRTOS, which is not suitable for Pico SDK or MicroPython."

But it turned out that MicroPython already had a wireless LAN driver for a similar Infineon chip. "We were able to take that code and extend it to use this new SPI method," reveals Liam. That provided a base for the MicroPython driver and a Pico SDK driver. "So that's what we're launching with," says Liam.

"It's implicitly muxed," explains Dominic. When either the RP2040 or the Infinion 43439 aren't driving the pins then other elements have priority. "Muxing is done with some resistors," he tells us.

> I felt it was important to try and make sure they have the existing I/O

"There are some nice MicroPython libraries you can use, like requests. So if you want to talk to a REST API, it's quite easy to do that in MicroPython. You just say 'connect to my access point' and once connected, you can send and receive data.

"The main thing we've been using at the moment is iPerf (**iperf.fr**), which is network speed testing. It's been good for us to see how well the device performs with different code and also stress it as much as possible and make sure it's reliable."

The potential for wireless projects is vast. You can do "anything from remote temperature sensing, logging, power usage around the house," says Dominic: "anything, really".

"You can even run small web servers on them," adds Liam. "So if you want a basic web page where you click a button to set or get a GPIO pin, then that's possible."

What about Pico WH?

Pico WH is a Pico W with a plastic header attached to make prototyping and connecting to accessories easier. This is in development because it requires some further design changes. The plastic header shrouds hold the sides, and that takes up even more space. And because Pico WH has a different debug connector – vertical rather than horizontal – this will need redesigning to work with the new serial debug pins on Pico W.

"Pico WH is on its way, but is not available yet," Dominic tells us.

Exploring electronics with a Pico Breadboard Kit

Raspberry Pi Pico W brings physical computing and the internet together and it's never been easier. Let's learn the basics by making a weather indicator

MAKER

PJ Evans

PJ is a writer, software engineer and general tinkerer. He can currently be found replacing all his old microcontroller projects with Raspberry Pi Pico Ws.

@MrPJEvans

If the new Raspberry Pi Pico W's wireless LAN capability has got you eager to start making but you're not sure where to start, you're in the right place. We're going to take a step-by-step look at simple components for inputs and outputs, connect them to Pico W, and then fetch data from the internet and display them. We're going to do this without any of the difficulty of soldering, or even handling components, by using the SB Components Pico Breadboard kit. This PCB (printed circuit board) comes pre-populated with buttons and LEDs to make your introduction to electronics as simple as possible.

01 Pico preparation

A great facet of Raspberry Pi Pico W is its support for different languages. By uploading different 'firmwares' (low-level code that translates for the RP2040 CPU), we can write programs in different ways. We're using MicroPython, a microcontroller flavour of Python, which greatly simplifies writing code for Pico W. It also includes everything we need to connect to the internet. To download and install the latest version of the MicroPython firmware, follow the instructions here: **magpi.cc/micropython**.

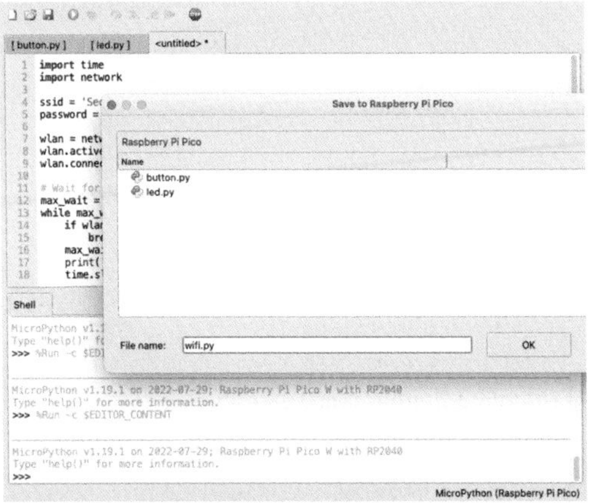

▲ You can write code as files and upload them directly to Pico W. To have code run instantly when Pico W is plugged in, just name it **main.py**

You'll Need

- SB Components Pico Breadboard Kit **magpi.cc/picobreadboardkit**
- 2 × 20-pin headers (if not already there) **magpi.cc/picoheaders**
- 10 × Female-female jumper cables **magpi.cc/jerkyjunior**
- Micro-USB to USB cable suitable for your computer

02 Get ahead

To connect to the Pico Breadboard Kit, you will need to have headers soldered onto your Pico. If you do not have a Raspberry Pi Pico H (H for headers), then you need to buy a header kit and solder them on yourself. These need to be facing downward so the smaller part of the header is poking through the top of Pico's PCB on the RP2040 chip side. If soldering yourself, remember to be careful and start by soldering each end pin of the header block, then check everything is level. If not, you can melt the solder to move them into place, then solder all the ones in between.

03 Choosing a development environment

To write code for Raspberry Pi Pico W, you need to use a computer. Nearly any modern operating system will do, including Raspberry Pi OS. It's possible to write code using a simple text editor, but it's a lot easier (and faster) to use an IDE (integrated development environment). Don't be put off by the fancy name – this is a text editor that understands what a Pico W is and can help transfer programs. On Raspberry Pi OS we recommend Thonny, but you can also use Visual Studio Code with the Pico-Go extension. More info: **magpi.cc/gettingstartedpico**.

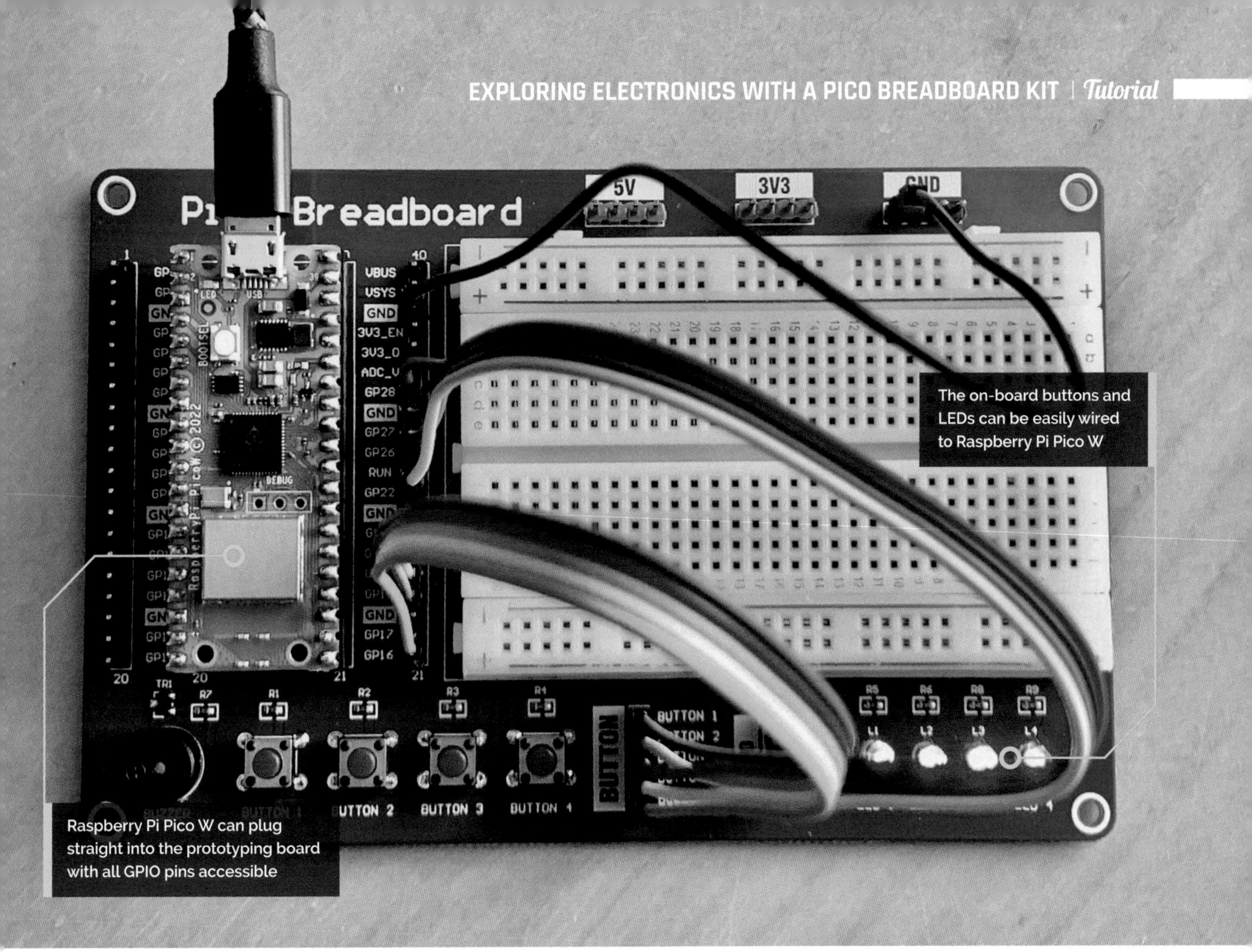

The on-board buttons and LEDs can be easily wired to Raspberry Pi Pico W

Raspberry Pi Pico W can plug straight into the prototyping board with all GPIO pins accessible

04 Testing time

It's time to check everything is working. Connect your Raspberry Pi Pico W to your computer using a USB cable. Open up Thonny and look in the bottom right-hand corner. It should say something like 'Python 3.7.9'. Click on this and, if Pico W has been recognised, you'll see 'MicroPython (Raspberry Pi Pico)' as an option. Select this and you'll see a welcome message on the bottom half of the screen. Click to the right of the '>>>' prompt and type:

```
print('Hello')
```

...followed by pressing **RETURN**. If you see 'Hello' displayed in response, you've just run your first program on your Pico W!

05 Understanding LEDs

Have a look at the SB Components prototype board. On the bottom-right are four LEDs (light-emitting diodes). These are one of the most common components used when beginning electronics, as we can make them light up and that's cool! LEDs can be a little naughty and draw too much current if left unchecked, causing permanent damage. To stop this happening, we need an in-line resistor to limit the flow of current. See the little black and silver squares above the LEDs? They are 330 Ω resistors already in place that are perfect for the job, so we can wire everything up without worrying about damaging our LEDs.

> We are using the RP2040's built-in 'pull-down' resistor circuit which solves this problem

06 Understanding buttons

On the left-hand side of the prototyping board are four buttons (and a buzzer, but we'll get to that). Each button creates a circuit when the button is pressed down. By wiring these to your Raspberry Pi Pico W, you can detect when the button is pressed. Buttons can be tricky for a microcontroller to handle, as the input is so sensitive it can give inaccurate readings; you can even trigger it by putting your finger next to it. To prevent this, we are using the RP2040's built-in 'pull-down' resistor circuit which solves this problem. When it comes to coding, you'll see how we make use of it.

07 Get wired

It's time to assemble our circuit. Carefully insert the disconnected Raspberry Pi Pico W into the socket on the prototype board, with the USB end at the top. The LEDs and buttons connect to Pico W using jumper cables. Between the buttons and the LEDs you'll see two yellow sets of headers, clearly labelled. The jumper cables need to run from these to the GPIO pins on Pico W. If you're wondering where the ground connection is, look at the top-right of the board. The GND header must have one wire connected to any of the GND pins on Pico W. Follow the wiring table (overleaf) carefully.

08 Light up the LEDs

Having checked all your wiring carefully, connect Raspberry Pi Pico W to your computer. In Thonny, type the **leds.py** code listing into the upper window, and then click the 'Run' icon. When prompted, ask to save it on the Pico W and name it **leds.py**. The code will now be uploaded to Pico W. Do you see the LEDs coming on one-by-one? The code starts by telling Pico W which GPIO pins are connected and what they are for (in this case – output). Then we go into a loop: toggling the state of each pin, then waiting a second. If you want, have a play with the sequence or see if you can change timings.

09 Push the button

To check the buttons, create a copy of the file you created in Step 8, and call it **buttons.py**. Remove the block starting `while True:`, then add the contents of the **buttons.py** code listing. Save and run the code. Try pressing the buttons one by one. Each one should now toggle its equivalent LED. This code uses 'event' or 'interrupt' handlers, blocks of code that run when a GPIO pin changes state. When a button is pressed, the code runs and changes the state of the LED. This is a fundamental part of physical computing. You are taking external input (the button) and creating output (the LED).

10 Simon says

You've now seen how Raspberry Pi Pico W can use code to respond to inputs by

Top Tip

Keep it in order

LEDs or buttons not working in the right order? Check the wiring, it's really easy to get things the wrong way around.

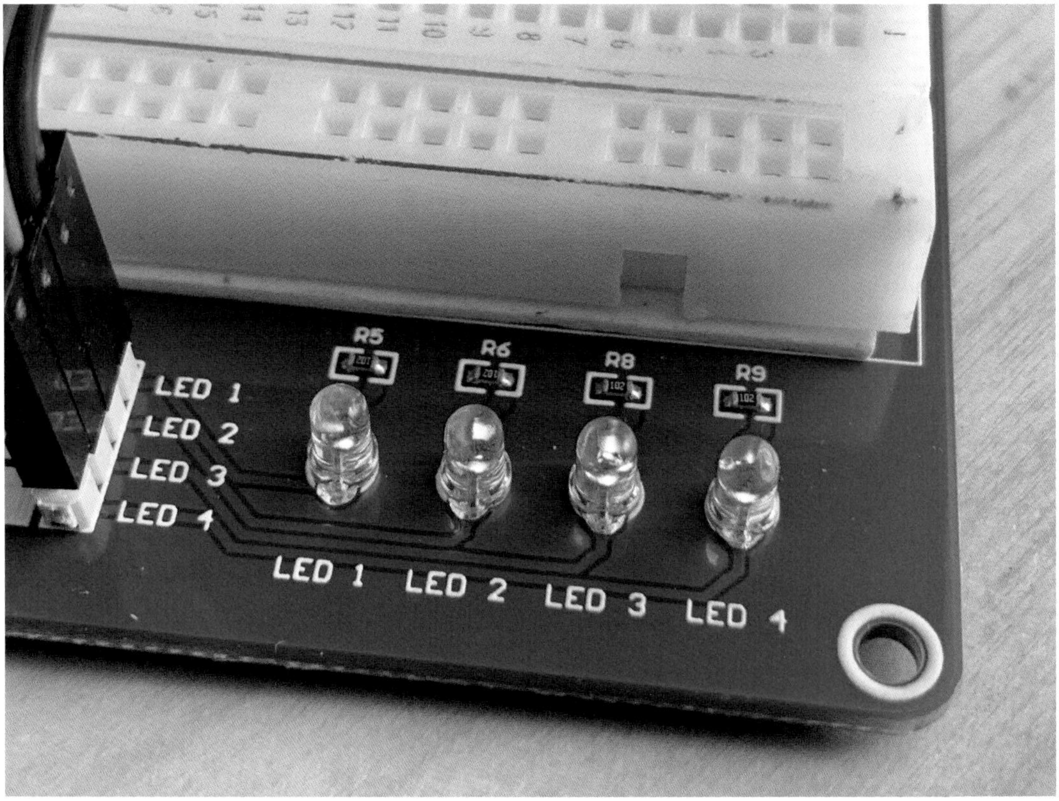

LEDs require in-line resistors to prevent them drawing too much current. Here, those resistors are already provided

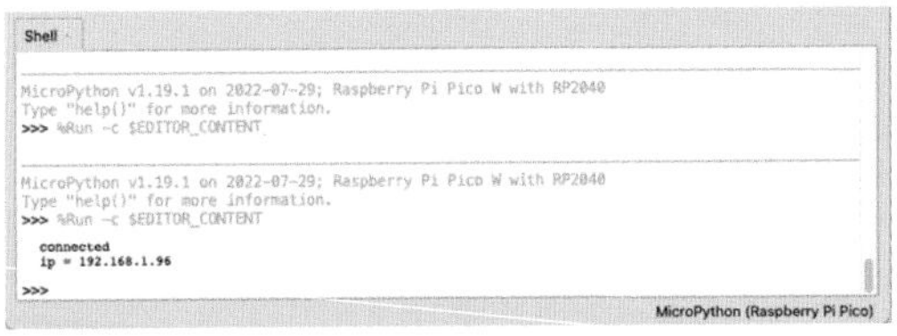

▲ With a few lines of code, Pico W can connect to your wireless network and then to the internet

creating outputs. We could have wired the buttons directly to the LEDs to create a similar effect (and using the breadboard, you can try that!), but Pico W adds logic that would be hard to implement in raw circuitry alone. To demonstrate this, download **memory.py** from **magpi.cc/memorypy**. This is an extension of the code that turns our button script into a memory game. Run the code on your Pico W, and see if you can remember the sequence of LEDs by playing them back on the buttons. Don't forget to review the code and see how it works!

> "If you're wondering where the ground is, look at the top-right of the board"

11 Get online

Now we have built our circuit, tested it, and played a game, let's look at what makes Raspberry Pi Pico W so special. For our weather project, we need to connect to the internet, so let's start with that. Create a new file called **wifi.py** and add the contents of the **wifi.py** listing (overleaf). Replace the `ssid` and `password` values with those for your own network. Now run the code on your Pico W using Thonny. Watch the console output and within a few seconds you should get an IP address announcement, meaning you're on the internet!

12 A key step

We're going to get some weather info to display on our prototype board. We'll use **openweathermap.org** to supply information using an API call. This is just like getting a web page, except the information is returned in a way computers can easily understand (in this case JavaScript Object Notation, or JSON). Sign up for a free account on the site and, once logged in, go to

leds.py

Language: **Python 3**

DOWNLOAD THE FULL CODE: **magpi.cc/ledspy**

```
from machine import Pin
import utime

# Make sure these are the pins connected to your LEDs!
leds = {
    1: Pin(28, Pin.OUT),
    2: Pin(27, Pin.OUT),
    3: Pin(26, Pin.OUT),
    4: Pin(22, Pin.OUT),
}

# Loop through the LEDs toggling each one then sleeping a second
while True:
    for i, (k, led) in enumerate(leds.items()):
        led.toggle()
        utime.sleep_ms(1000)
```

buttons.py

Language: **Python 3**

DOWNLOAD THE FULL CODE: **magpi.cc/buttonspy**

```
# Remove the while True: block and replace with this
def button_handler(pin):
    button_pressed = int(str(pin)[4:6]) - 17
    print(str(button_pressed))
    leds[button_pressed].toggle()

for gpio_number in range(18, 22):
    button = Pin(gpio_number, Pin.IN, Pin.PULL_DOWN)
    button.irq(trigger=Pin.IRQ_RISING, handler=
button_handler)
```

'API Keys'. One will have already been created for you (although it can take an hour or two to start working). Think of this as a password allowing you to access the service. Take a copy – you're going to need it soon.

13 Talk about the weather

We're going to make a request for the current weather. Download **weather_1.py** from

▲ These handy jumper cables make connecting the LEDs and buttons to Pico W easy and safe

▼ Here are the connections you need to make between Pico W and the button and LED connectors. Be careful and don't forget the GND!

magpi.cc/weather1py, then replace the three new variable values at the top with the API from the previous step and your desired latitude and longitude. Don't know these? Just enter 'lat and long for town' in a search engine, and you'll get the answer. The ones in the code are for the Raspberry Pi Foundation in Cambridge. Run the code on your Raspberry Pi Pico W as before, and watch the console output. Here, we use the urequest library to request information from the API server.

	Button	LED	GND 1-4
GP28		1	
GP27		2	
GP26		3	
GP22		4	
GP21	1		
GP20	2		
GP19	3		
GP18	4		
GND			Any

14 Hooking up the buttons

We've got four buttons, so let's pick out four key pieces of information. Each time we press a button, a request will be made to the API and the code will extract a useful piece of data from the request. We're going to ask for temperature, windspeed, rain, and air quality index. This code is a little longer, so download **weather_2.py** from **magpi.cc/weather2py** and transfer it to your Raspberry Pi Pico W as before. Run the code and press each button. Watch the output in the console as you press the buttons.

> "Enter 'lat and long for town' in a search engine, and you'll get the answer"

15 Lightening

We don't have a screen (although you can add one if you want!), just four lights to show the data. What we'll do is divide the results into ranges and light the appropriate amount of lights. For instance, If it's really hot, all four LEDs will illuminate. Download **weather_3.py** from **magpi.cc/weather3py** and run it. Raspberry Pi

You could use a standard breadboard and components, but the Pico Breadboard Kit makes it all a lot easier

Pico W will download the JSON data and extract our four data points and display them using the LEDs. Feel free to change the ranges if you wish.

16 Buzzin'

Calls to the internet can never be fully relied on to succeed. There are many things that can go wrong, from your internet connection being down to the API server having problems. We can catch these errors and signal to the user that there's a problem. The next version of our code (**weather_4.py** from **magpi.cc/weather4py**) creates a short buzz on successful calls and a longer buzz if something went wrong. Keep getting errors? More details will be logged to the console.

17 Put it all together

Let's bring the buttons and the LEDs together. Download our final code version, **weather_5.py** from **magpi.cc/weather5py**, and run it up as before. Now when you press each button, you can get an idea of whether it's raining, sunny, hot, or windy! Take some time to walk through the code to see how we hunt through the data, and see what changes you can make! If you would like to run this independently without a computer attached, just rename this file to **main.py**. Any file of that name will run automatically when power has been applied to Raspberry Pi Pico W.

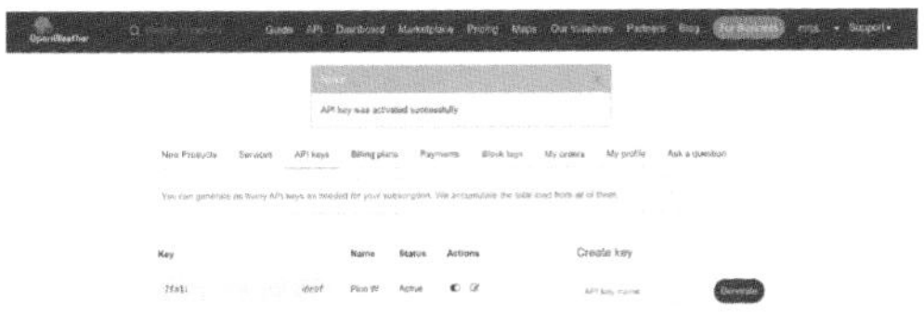

To get the weather data you need an API key. You'll find it on the OpenWeatherMap user page like this

18 Make it your own

In this tutorial we've learnt how to control LEDs, listen for button presses, and combine those with internet data. Feel free to alter the code to show different things. Maybe you could periodically check the API and sound the buzzer when it's raining? This is just the beginning! For such a low-cost device, the capabilities of Raspberry Pi Pico W go much further than switches and lights. You can add all kinds of sensors, screens, and even motors with the right kit. Get an electronics kit and use the breadboard to add more features. There are endless tutorials out there to help you along. Be curious and have fun!

Top Tip

More data

OpenWeatherMap offers different APIs, many for free, so it's worth exploring what other data you could get, such as UV warnings.

wifi.py

Language: **Python 3**

DOWNLOAD THE FULL CODE: **magpi.cc/wifipy**

```
# Based on code by Pete Gallagher
# https://www.petecodes.co.uk/
import time
import network

ssid = "<Your Wifi Network Name>"
password = "<Your Wifi Password>"

wlan = network.WLAN(network.STA_IF)
wlan.active(True)
wlan.connect(ssid, password)

# Wait for connect or fail
max_wait = 10
while max_wait > 0:
    if wlan.status() < 0 or wlan.status() >= 3:
        break
    max_wait -= 1
    print('Waiting for connection...')
    time.sleep(1)

# Handle connection error
if wlan.status() != 3:
    raise RuntimeError('Network connection failed')
else:
    print('Connected')
    status = wlan.ifconfig()
    print( 'IP Address = ' + status[0] )

# Important to tidy up the connection
wlan.disconnect()
```

Reviews

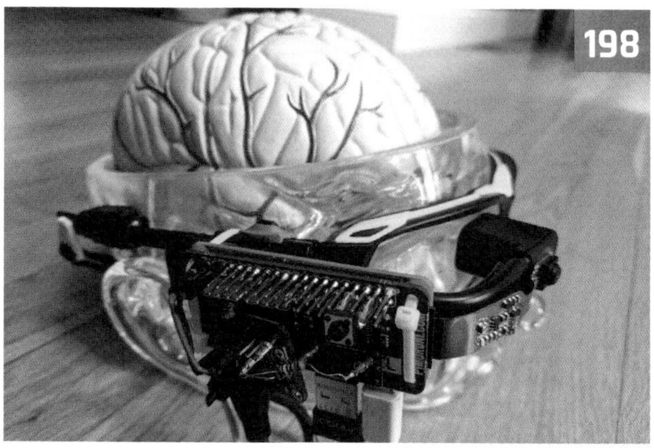

200

184

Grove Starter Kit for Raspberry Pi Pico

SPECS

SIZE: 56 mm × 56 mm

PORTS: 3 × analogue, 3 × digital, 2 × UART, 2 × I2C, 1 × SPI

OPERATING VOLTAGE: 3V3 / 5V

▶ Grove ▶ **magpi.cc/grovestarter** ▶ £TBA / $43

Explore the world of Raspberry Pi Pico electronics with this click-and-play kit. By **Lucy Hattersley**

Exploring the world of electronics is one of the fundamental joys of using a microcontroller such as Raspberry Pi Pico (magpi.cc/pico). The GPIO pins on Pico can be wired up to an array of sensors, buttons, LEDs, and displays enabling a wide array of projects.

At the heart of the Grove Starter Kit for Raspberry Pi Pico is a Grove Shield ($4, **magpi.cc/groveshield**). This board has ten different Grove Connectors: three analogue ports, two digital ports, three UART ports, and four I2C ports. It also has a 3V/5V power switch to adjust the voltage output, and a six-pin SPI interface.

Raspberry Pi Pico slots into the header on the Grove Shield and Grove parts are snapped straight into the white Grove connectors. This enables you to prototype projects quickly and easily.

You will need to have soldered pins on to your Pico (or you can pick up a pre-soldered Pico, **magpi.cc/picosoldered**). But from that point on, you don't need to solder parts or figure out circuits and jumper wires. You just connect the Grove part to the Grove Shield using one of the included Grove Universal 4-Pin cables (**magpi.cc/grovecable**). Grove itself says this "simplifies the learning system, but not to the point where it becomes dumbed down".

Each Grove cable has four wires: typically one for power, another for ground, and two for input and output (the exact nature of each wire depends on the part it's connected to; you can read more at **magpi.cc/groveinterface**).

This does, indeed, make it extremely easy to hook up components to Raspberry Pi Pico. And to that

▶ The Grove Shield connected to a 16×2 LCD and two sensors. Here the display is programmed to output light and sound values

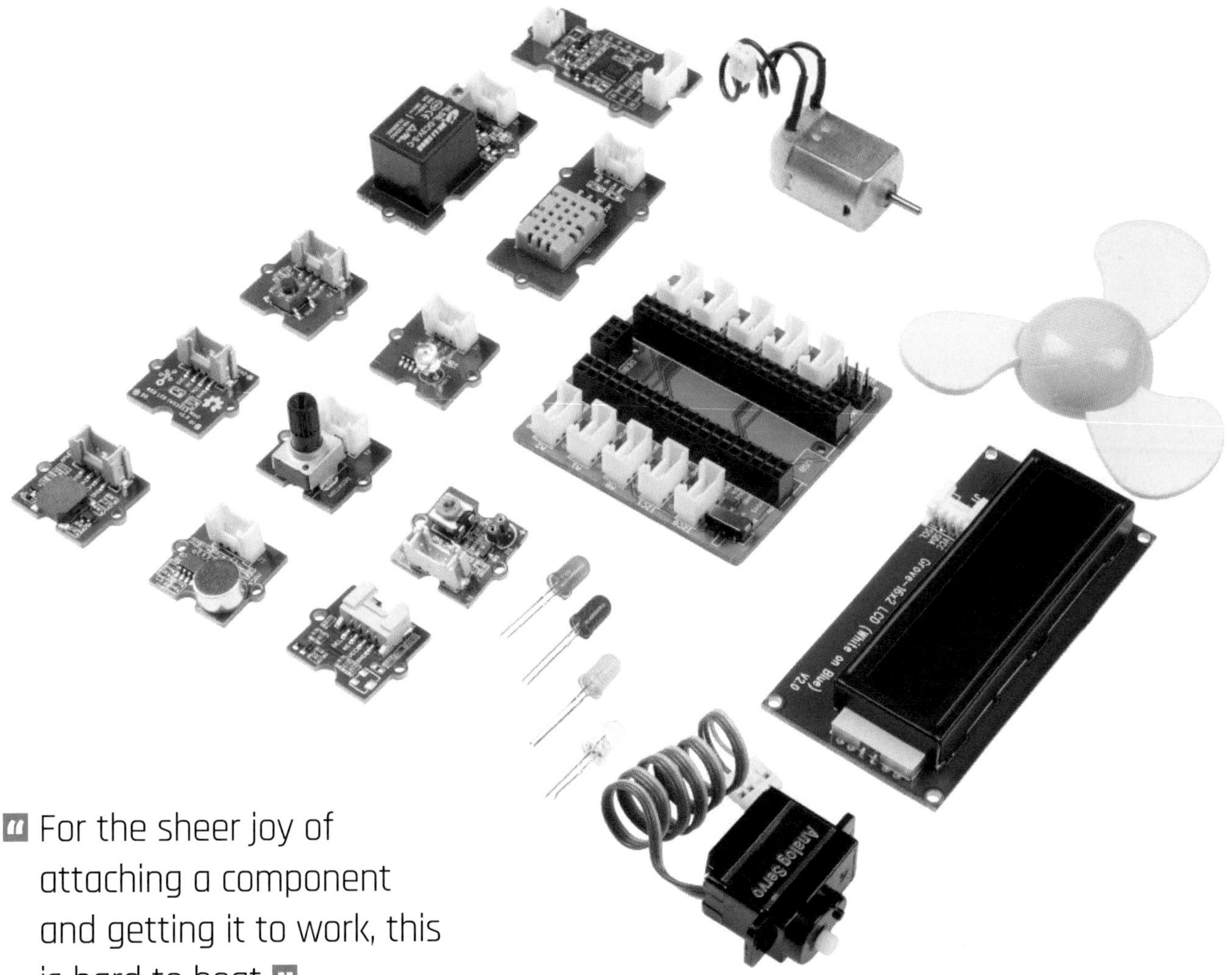

▲ The Grove Starter Kit contains a range of Grove components and the Grove Shield

> For the sheer joy of attaching a component and getting it to work, this is hard to beat

end, the kit comes with a wide range of parts to play with. There is an LED pack, RGB LED display, light sensor, sound sensor, rotary angle sensor, temperature and humidity sensor, passive buzzer, button, servo, mini fan (with DC motor), relay, and a 16×2 character LCD.

Learning curve

The Grove Shield for Pi Pico wiki page has a range of tutorials that use the parts found in the kit (**magpi.cc/groveshieldwiki**).

The tutorials use all the parts found in the Starter Pack and give a good overview of what you can do. You typically need to download a Python module for each part. And analysing the code will give you a good overview of what each component can do. We programmed the LCD to respond to light and sound; a fan and servo movement detector; and played around with lights, buttons, and the relay.

Thanks to the Python support files, introductory wiki tutorials, and the click-and-play nature of the kit, it is ridiculously easy to move from having an idea to getting it working.

There is an argument that replacing the pure jumper wires and breadboard with a custom connector removes part of the learning curve. And it's hard to take a prototyped circuit and wire components directly to Pico so you remain attached to the Grove Shield and its ecosystem of parts. But, for the sheer joy of attaching a component and getting it to work, this is hard to beat.

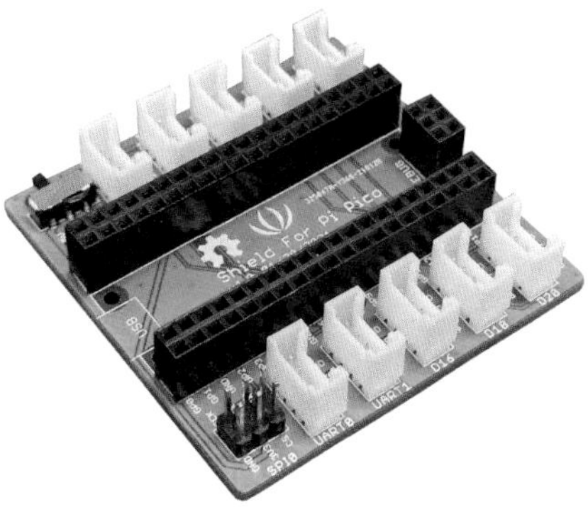

▲ The Grove Shield for Raspberry Pi Pico has a range of ports that make it easy to click and connect components

Verdict

It offers a painless way to attach components to Raspberry Pi Pico and play around with electronics. There's a huge range of components, and the documentation is good.

9/10

THine Cable **Extension Kit**

▶ THine ▶ **magpi.cc/cableextender** ▶ £51 / $59

Connect an HQ Camera to Raspberry Pi via an extremely long Ethernet cable. **Lucy Hattersley** takes a look at this long shooter solution

SPECS

BOARDS: Transmitter Board with THine THCV241A MIPI CSI-2 to V-by-One HS Serializer; Receiver Board with THine THCV242 V-by-One HS to MIPI CSI-2 Deserializer

COMPATIBILITY: Camera Module V1.3 (limited support); Camera Module 2.1; HQ Camera

DIMENSIONS: TX board: 38×25 mm; RX board: 65×56 mm

THine Cable Extension is a kit for Raspberry Pi that swaps out the CSI cable (Camera Serial Interface) for a LAN/Ethernet cable. This enables you to dramatically increase the length of the cable that can be used, positioning a Camera Module up to 20 metres away from Raspberry Pi.

The kit comes in two parts: the Transmitter Board attaches to your Raspberry Pi Camera Module / HQ Camera; the Receiver Board connects to Raspberry Pi via the GPIO pin header.

Each board features a CSI socket and RJ45 (Ethernet) socket and the two are connected via a CAT5e or higher LAN/Ethernet cable. The system is designed to be 'plug and play', so all you need to do is hook everything up and use the Camera Module as normal.

Japanese company THine's V-by-One HS Serializer / Deserializer technology provides a "high speed data link solution for MIPI CSI-2, camera serial interface". Learn more on THine's website: **magpi.cc/THCV241A**.

The Cable Extension kit comes supplied with the two boards, a 2-metre LAN cable and the ribbon flex cables, along with mounting screws. Our test kit also came with a Raspberry Pi 4, HQ Camera, and a 5 m Elecom CAT6 Ethernet cable.

We took everything apart and set it back up with a fresh installation of Raspberry Pi OS, then followed the quick-start guide (**magpi.cc/cableextensionqs**).

▶ The two THine Cable Extender boards convert a Camera Module's CSI cable into a potentially much longer LAN/Ethernet cable

Snap happy

Setup was hassle free and it 'plugged and played' as outlined by THine. There was no need for any additional software installation.

One thing of note: this is not a network-enabled solution. Instead, the LAN/ Ethernet cable is designed to run directly from Raspberry Pi to Camera Module.

We tested it around the house and ran the cable out into our garden where we kept an eye on a bird feeder from afar.

And 20 metres is a massive upgrade from the 20 cm CSI cable included with the Camera

THine Cable Extender is ideal for running a Raspberry Pi HQ Camera outdoors while Raspberry Pi remains safely out of the elements

> Works exactly as outlined and performs a useful function admirably well

Module. We checked with Raspberry Pi, and there is a limit with CSI before you lose signal integrity. There's no guarantee that a CSI cable longer than the 20 cm one supplied will work.

So this is a big upgrade in effective distance. The Cable Extension Datasheet (**magpi.cc/cableextensiondatasheet**) has more info on cables tested up to 20 m with this kit.

It can be a mild trouble to fine-tune the HQ Camera Module's adjustment rings with the Raspberry Pi screen so far away. Apart from that, we struggle to find any downside. Cable Extension Kit works exactly as outlined and performs a useful function admirably well.

There are many use cases where it is beneficial to keep the camera and Raspberry Pi some distance apart, especially when filming in a hostile environment. THine outlines a project where they set up an outdoor PiKon 3D-printed telescope (**pikonic.com**) and used it at night during a freezing Chicago winter: **magpi.cc/cableextensionpikon**).

THine Cable Extension Kit is a niche product for a niche use case. But the technology is clever and it works exactly as outlined with no fuss. If you have a desire to place your Camera Module some distance away from Raspberry Pi, then this is the way to do it.

Verdict

A breeze to set up, and works exactly as outlined by THine. The V-by-One HS technology is interesting and this is the way to go about connecting a Raspberry Pi Camera Module over a long distance.

9/10

Marty the **Robot V2**

▶ Robotical ▶ **robotical.io** ▶ From £377 / $420

We take this two-legged robot for a walk and wave. By **Lucy Hattersley**

SPECS

PROGRAMMING LANGUAGES: Scratch, Python, JavaScript, ROS (Robotic Operating System), OpenAPI

CONNECTIVITY: I2C, Bluetooth, USB, serial, WiFi

BATTERY: 2600 mAh Li-ion rechargeable battery, 2-3 hours of use on full charge

SENSORS: Accelerometer, tilt sensor, motor current sensors, motion position sensing, fall detection, IR proximity, colour & line sensor. Optional extras: distance sensor, noise sensor, compass, temperature sensor

Marty the Robot V2 is a small, personable two-legged robot that uses nine different servo motors to move around. It's capable of a wide range of small movements, including walking in both directions, rotating, side-stepping, kicking, waving, and waggling its eyes.

All of this movement is controlled by a custom RIC (Robot Interface Controller) based upon an ESP32 microcontroller (**magpi.cc/martytechspecs**). It has built-in Bluetooth, wireless LAN, an accelerometer, and a speaker. It's powered by a rechargeable battery with a built-in USB-C charger.

With his big blue head and various accessories, Marty is a good-looking piece of kit, the sort of personable toy that children can make friends with. However, there's a lot to unpack here, surprisingly so for a device that looks very much at home on the shelf of a local toy store. There is a scalable level of complex interaction that can move right from key stage 2 (7-year-old and upwards) to the university level.

Right at the entry level, you don't need any computer at all. With an optional IR colour sensor (£19, **magpi.cc/martycolour**), Marty is controlled by placing coloured cards on the floor: green to move forward, purple to slide right, and so on.

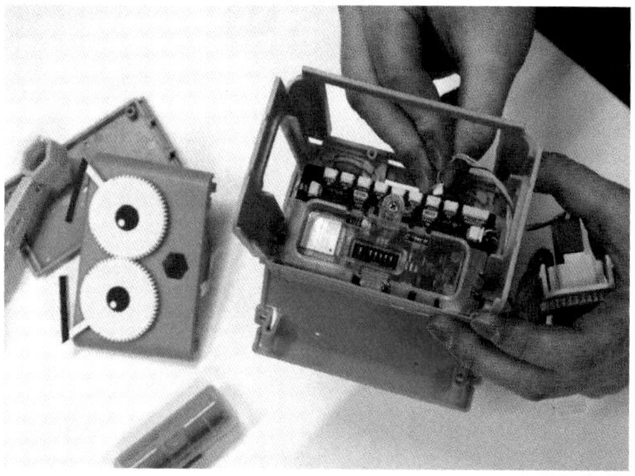

▶ Inside Marty is a RIC (Robot Interface Controller) that can connect to Raspberry Pi via a wireless network

The next stage of interaction is via an Android or iPhone app. This connects to Marty via Bluetooth and provides button controls (walk, rotate, and so forth). The app also has a built-in version of Scratch 3 that is packed with blocks to control Marty. So children can use Scratch to create programs for Marty and learn the basics of programming with a physical device (much more interactive than watching a sprite shuffle around the screen). To this end, there are a range of teacher guides and lesson plans available (**learn.martytherobot.com**).

Raspberry Pi connection

The app is also used to connect Marty to a local wireless LAN, which is where a Rtaspberry Pi can step in. Once on your network, you can move from

▼ Fitting Raspberry Pi inside Marty V2

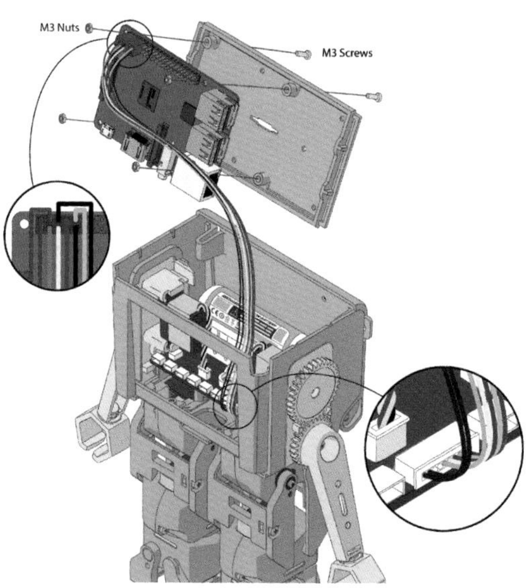

> We found Marty personable, fun to play with, and packed with features – some of which scale surprisingly high up

Scratch to Python and connect to Marty using the MartyPy library. For example, here was our program to make Marty dance:

```
from martypy import Marty
my_marty = Marty("wifi","192.168.0.45")
my_marty.dance()
```

There's a comprehensive Knowledge Base portal on Robotical's website (**magpi.cc/roboticalknowledge**) with example code, tutorials, and a Python function reference guide.

Marty does not – by default – contain a Raspberry Pi computer. Instead, you connect to the built-in RIC (Robot Interface Controller). However, you can install a Raspberry Pi inside the head unit and connect it to the RIC (**magpi.cc/martyraspberrypi**) using the supplied cable.

Putting a Raspberry Pi inside Marty gives it independence from the network and other computers or the app. There's also a decrease in latency time, so response time is faster. There's the tantalising possibility to expand Marty's abilities with voice interaction and image processing to make the robot more like a walking smart assistant.

Even if students never make it that far, learning code by moving a physical device around, rather than shuffling sprites around on the screen, is a powerful concept. And we found Marty personable, fun to play with, and packed with features – some of which scale surprisingly high up. The detailed documentation helps get you started and gives you direction.

Verdict

A great little robot that brings Scratch and Python code to life. Don't be fooled by the cute exterior: there's a lot of potential inside this robot.

8/10

pi-top [4] Robotics Kit and Expansion Plate

▶ pi-top ▶ **magpi.cc/pitoprobot** ▶ From £187 / $200

Build a range of robots with aluminium plates and reusable rivets, then control them with Raspberry Pi and pi-top [4] DIY Edition case. This kit clicks with **Lucy Hattersley**

SPECS

MOTORS: 2 × 12 V high-torque geared motors with Hall effect sensor tachometers; 2 × modular servo motors

SENSORS: HD 720p wide-angle camera module; 1 × ultrasonic sensor

COMPONENTS: Chassis interface plate; 25 mm Durable omni-directional castor wheel; 2 × 74 mm wheels with all-terrain tyres; 50+ aluminium plate construction pieces; 200+ nylon quick-build rivets

EXPANSION PLATE: Accelerometer, gyroscope, and magnetometer; 4 × 6P 12 V DC motor ports; 4 × 3P servo motor ports; 2 × USB 2.0 ports; camera (CSI) and display (DSI) ports; 8 × digital and 4 × analogue sensor ports

Pi-top [4] Robotics Kit is a long-awaited robot that complements the pi-top [4] DIY Edition case. It promises to be a sturdy and versatile building platform, with competent software and good educational chops.

And it delivers. Inside the box is a series of aluminium plates and plastic rivets that act a little like LEGO Bricks meet Meccano. Two rivet-compatible servo motors and two encoder DC motors, plus a webcam and ultrasonic distance sensor. Everything you need to build a wheeled robot that can see and sense the world around it.

A plan comes together

We reviewed the pi-top [4] DIY Edition (**magpi.cc/pitop4review**) back in *The MagPi* issue 99 (**magpi.cc/99**). At the time, we were impressed with the build quality, but noted the oddity of the built-in battery and 128×64 OLED display.

Once clipped into the Expansion Plate on the robot, it all makes sense. The pi-top [4] case powers both Raspberry Pi and the motors in the robot; the OLED display provides feedback on the IP address and remaining battery level (we got around two hours of use). The kit tested comes with an Expansion Plate that connects to the bottom of the pi-top [4] DIY Edition case and breaks it out into several control ports.

Making robots

Thanks to the rivet system, you can get creative with your robots. Three designs are included: Alex, a regular wheeled robot with a pan-tilt mechanism for the camera and ultrasonic sensor; Bobbie uses the servos to control two ping-pong ball grabbing pincers; Prax is angled in a vertical position and the servos create a moving face for a desk-based interactive assistant. Instructions for all three builds are available as a PDF download from the pi-top Start website (**pi-top.com/start**) and it's a good place to get an overview of how the builds work.

Going further

Each build took us around an hour. When the build is complete, you attach a pi-top [4] DIY Edition to

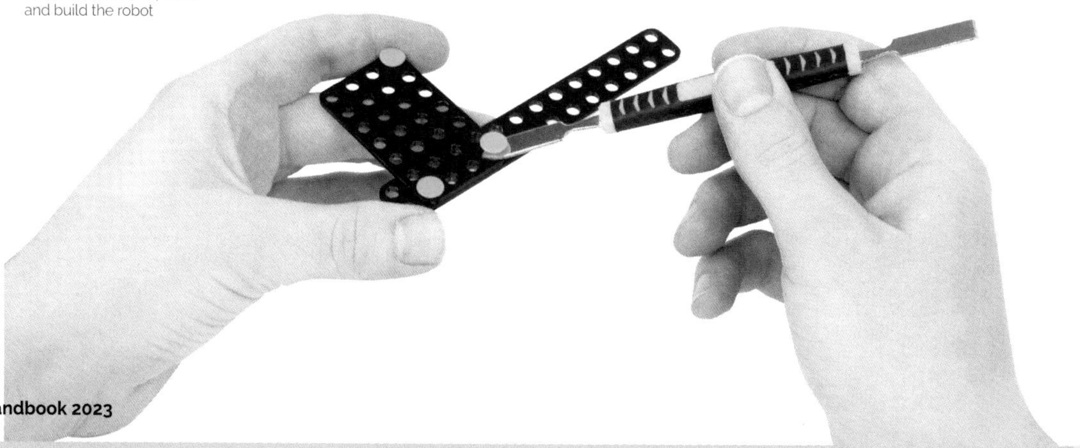

▼ The innovative rivet system enables you to connect aluminum plates and build the robot

The "Prax" design uses the components to build an interactive desktop assistant

The pi-top Robotics Kit designs are sturdy. The "Bobbie" design uses servo motors as two pincer arms

> Superbly designed with a clever rivet construction system and seamless integration with pi-topOS and Further courses

the Expansion Plate and insert the cables to connect the electronic parts to the Expansion Plate.

Here is where pi-top [4]'s on-board battery and OLED display spring into useful action, providing network information that you can use to quickly SSH into the robot.

Pi-top has clearly put a lot of elbow-grease into its pi-topOS and its Further 2.0 system (**magpi.cc/further**). The pi-topOS ensures elements like SSH are enabled by default; along with baked-in support for the hardware components.

The pi-top [4] DIY Edition sits on top of the chassis as the brains and battery of the robot

The Further website has demonstration code to follow, and you can SSH directly to your robot from the Further website and run the interactive code examples directly from the web. Students can log in separately and collaborate on code and practise programming techniques together. We enjoyed the coding courses and challenges, and integration with OpenCV for object and face recognition.

Moving beyond the Further educational port, you can code pi-top [4] Robotics Kit directly with the pi-top Python SDK (**magpi.cc/pitopsdk**) and there is support for ROS (**magpi.cc/pitopros**) and Microsoft's .NET (**magpi.cc/pitopnet**).

There's a lot here. First of all, the whole kit is not particularly cheap if you include the price of the pi-top [4] DIY Edition and a Raspberry Pi 4. It is, however, superbly designed with a clever rivet construction system and seamless integration with pi-topOS and Further courses. All in all, pi-top [4] Robotics Kit is plain nice to use, and will sit neatly into an education environment. Good job.

Verdict

Very high-quality robotic education kit, with an innovative rivet construction mechanism and well-thought-out software. It's been worth the wait.

10/10

Pico Wireless Pack

▶ Pimoroni ▶ magpi.cc/picowireless ▶ £12 / $14

Add wireless networking to your Pico projects. By **Phil King**

SPECS

PROCESSOR: ESP32-WROOM-32E module

CONNECTIVITY: WiFi 802.11b/g/n, 2.4GHz

FEATURES: RGB LED, microSD card slot, tactile button

While Raspberry Pi Pico is an inexpensive yet powerful microcontroller, one drawback for IoT projects is the lack of built-in wireless LAN.

That's where the Pico Wireless Pack comes in, enabling you to connect Pico to your WiFi network over the 2.4GHz band. The Pack's wireless connectivity is provided by an on-board ESP32 MCU module via SPI.

> The obvious downside is that it blocks access to Pico's GPIO pins

In addition, the board features a bright on-board RGB LED, tiny tactile switch, and a microSD slot for storing extra data, expanding greatly on Pico's 2MB of flash storage.

To connect your Pico, you'll need to have male headers soldered on, then you can just plug it into the female headers of the Wireless Pack, so it and Pico are back to back.

The obvious downside is that it blocks access to Pico's GPIO pins. So, to connect sensors or other components, you'll need a stacking header on Pico or a breakout board such as Pimoroni's Pico Omnibus. Still, 14 GPIO pins are left unused; plenty for most projects.

Software libraries

Libraries for C++ and MicroPython can be downloaded from Pimoroni's GitHub repo. The latest version of Pimoroni's MicroPython UF2 firmware contains the necessary drivers, so you'll need to flash Pico with it.

To help you get to grips with the picowireless library, a couple of MicroPython examples are provided. One creates a simple web server enabling you set the red, green, and blue values of the on-board LED. The other demonstrates connecting to an API, and lets you control the LED's colour using #cheerlights tweets.

Both examples are pretty verbose, with some complex HTTP request details, so are hard to follow unless you know your networking, but you could easily adapt them for your own projects.

Alternatively, you can code with CircuitPython, using Adafruit's CircuitPython and ESP32 libraries, by adjusting a few pin numbers in their code examples.

▶ Adding wireless connectivity to Pico, the Wireless Pack also features an RGB LED, tactile button, and microSD slot for extra storage

Verdict

A simple way to add wireless connectivity to Pico, although you'll need to break out the GPIO pins for IoT projects.

8/10

HyperPixel 2.1 **Round**

▶ Pimoroni ▶ **magpi.cc/hyperpixel2r** ▶ £48 / $56

This high-quality circular display doesn't cut any corners. By **Phil King**

SPECS

DISPLAY: 2.1" IPS, 480×480 pixels, 18-bit colour, 60 fps frame rate

TOUCH INTERFACE: Capacitive touchscreen with multi-touch

BREAKOUT: Five-pin breakout header using alternate I2C

01

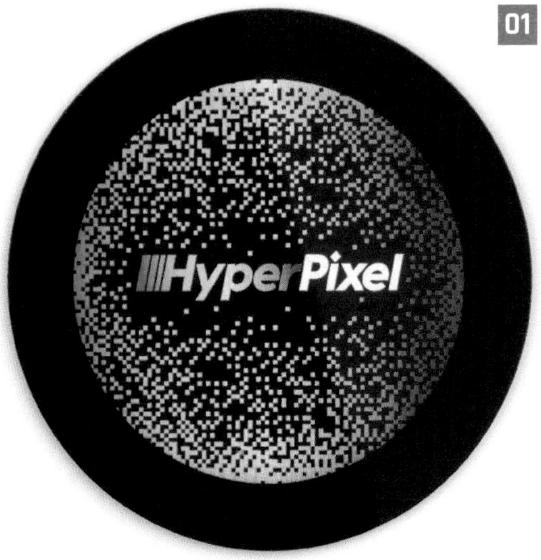

02

▶ 01. The screen is very vivid and can be oriented how you like – a terminal command is used to rotate the display output

▶ 02. A Raspberry Pi Zero fits neatly on the rear, which also features a five-pin breakout header (obscured)

Circular displays are all the rage in items like smart thermostats and high-end coffee machines. The HyperPixel 2.1 Round is the first one available for Raspberry Pi.

Essentially, it's a square 480×480 pixel display minus the corners. The bezel is a little wide at around 9 mm, but the touchscreen display itself is impressive. Like the standard HyperPixel, it uses a high-speed DPI interface, allowing it to shift five times more pixel data than the SPI interface typically used for other small screens.

A Raspberry Pi Zero fits neatly onto the rear, although you'll need to be careful when mounting it not to crack the screen – it's best to put it face down on a soft surface. Short standoffs (not supplied) can be use to secure it. If using a full-size Raspberry Pi, you'll need a GPIO booster header.

While the HyperPixel 2.1 Round uses all the GPIO pins, five breakout pins on the rear provide the option of connecting sensors via an alternate I2C interface.

Drive round

After installing drivers with terminal commands, rebooting sees Raspberry Pi desktop appear on the tiny display. Since the corners are missing, it's nigh on impossible to navigate, so you'll want to enable SSH access to issue further commands.

> “Like the standard HyperPixel, it uses a highspeed DPI interface”

The Python touch library is installed separately and contains a few code examples, including a clock and a colour wheel. Pygame is used for these, but since it doesn't support the non-standard 480×480 resolution, you have to alter Raspberry Pi's config.txt file to make it work – and add an extra part to the command when running code

At the first attempt, the code examples appeared glitchy on screen – an update of Pygame rectified this (see **magpi.cc/hyperpixel2rissue**) and it now runs correctly.

Videos run really smoothly at 60fps, although the aspect ratio is squished horizontally to fit the screen.

Verdict

If you really want a round display, maybe for a Halloween animated eyeball, this is an excellent option.

8/10

Keybow **2040**

▶ Pimoroni ▶ **magpi.cc/keybow** ▶ £50 / $57

An RP2040-based auxiliary keyboard for macros, samples, and streams. **Rob Zwetsloot** gives them a tickle

SPECS

KEYS: 16 RGB LED swappable keys in a 4×4 grid with mechanical, linear, and tactile key options

MICRO CONTROLLER: RP2040 with 2MB of flash memory

DIMENSIONS: 76 mm × 76 mm × 30 mm (L × W × H, including switches, keycaps and feet)

Ever since the release of Raspberry Pi Pico, we've been interested in its ability to power keyboards, both standard and custom. We've seen a couple of DIY projects doing just this, and Pimoroni even released an add-on with a soft keys (RGB Keypad: **magpi.cc/rgbkeypad**) for Pico at its launch. It was only a matter of time before the RP2040 – the chip that powers Pico – would be used on its own dedicated keypad with real keys. This inevitable product ended up being another Pimoroni product: Keybow 2040.

It comes with one of three sets of key switch types depending on your preference: mechanical (clicky-clacky), linear (silent), and tactile (silent with some resistance). All versions have an LED in the switch which is fully programmable, and the keys are standard sized so you can easily swap them out for others if you'd like. There's also a cheaper base kit for which you'll need to provide your own switches and keycaps.

▶ Each key is individually programmable, with customisable RGB LEDs as well

Lock on

Some construction is required, but it's very minimal – add the switches, sandwich the layers together, and then put the keys on. We were done in a couple minutes, at which point you can actually start using the Keybow as a number pad by plugging it in.

Unlike a standard Pico, the Keybow is connected via a USB-C cable to your computer of choice. There doesn't seem to be any technical

> "Much cheaper and more hackable than some of the commercial alternatives"

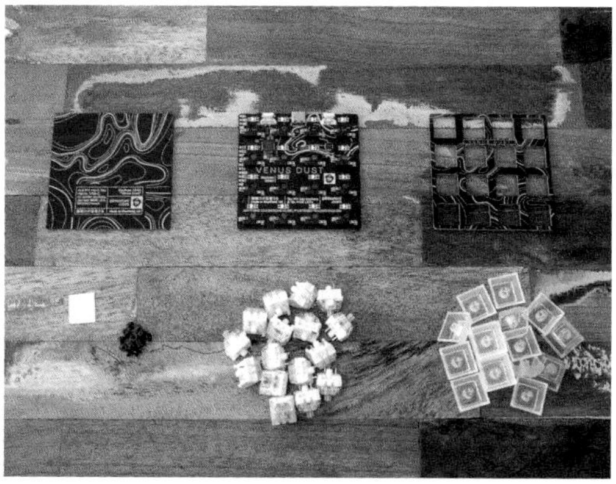

▲ Everything you get in the box – it seems a lot, but it's a quick build

It works just fine with a Raspberry Pi

advantage to having USB-C; however, it does mean the basic design has some forward compatibility for Pimoroni. Also, it's nice to have more stuff making use of USB-C. A couple of hardware buttons are also included: one to reset the Keybow, and one to hold down to do the boot selection if you want to tinker with the code.

Custom combo

The code itself is written in CircuitPython, which is derived from the MicroPython – both can be used on a regular Pico and they're largely similar as well. Whatever changes you need can be gleaned from the abundant example programs that are installed on Keybow 2040.

With little effort we were able to change the way the keys work – with constantly lit buttons that change when pressed, and passing along a specific combos of keys for when they're pressed rather than a single numpad key. Due to the way it loads scripts, you can immediately test out your edits, making for very quick prototyping and bug fixing.

We've been using a linear version for weeks now during streams as a custom stream deck, and it works perfectly with our custom setup (the code for which we'll drop on **magpi.cc/github**). It's sturdy, turns on quickly, the LEDs are bright, and we've had no noticeable lag on a button press. With the removable caps, you can add custom ones with specific labels like a mute icon or specific scene modes, etc., making it perfect for this, and much cheaper and more hackable than some of the commercial alternatives.

Verdict

A premium-feeling product that is easily customisable and priced much more fairly than less-open alternatives.

10/10

Midas 5in **TFT LCD**

▶ Midas Displays ▶ **magpi.cc/farnellmidas** ▶ £77 / $105

Turn Raspberry Pi into a professional touchscreen display system. By **Lucy Hattersley**

SPECS

DIMENSIONS AND ENVIRONMENT:
Module size: 120.70 × 75.80 × 23.20 mm; Operating temperature: -20°C ~ +70°C; Logic voltage: 5V

DISPLAY:
Display size: 5.0 inch; Resolution: 800×480; Orientation: Landscape; Brightness: 750 cd/m^2

INPUT/OUTPUT:
Touchscreen: CTP; Interface: HDMI; Pinout: 40-pin GPIO

Displays like this Midas 5in TFT LCD are used in a variety of industrial settings. Don't be surprised if you've already used one of these in a vending machine, EV charger, or information display.

Midas sent us a 5in IPS display for testing (part number: MDT0500D2IHC-HDMI, **magpi.cc/midas5in**). It is a small screen with a quirky 800×480 resolution and 5:3 aspect ratio. It also comes with a ten-point capacitive touchscreen and GPIO integration.

The screen is connected to the Raspberry Pi via the GPIO pins. A look at the datasheet (**magpi.cc/midas5indatasheet**) shows that GPIO17 is used to control the backlight, and power is passed through the GPIO pins (it is powered via a standard 5V supply connected to Raspberry Pi). A GPIO breakout on the rear of the device enables you to attach further electronic components or HATs.

The Midas 5in TFT LCD has the same resolution as the Official Raspberry Pi 7in Touchscreen Display (**magpi.cc/officialdisplay**). However, this device uses an HDMI connection for the display, and a micro-USB to USB-A cable to provide touchscreen feedback.

Our test unit came with a U-shaped MCIB-HDMI adapter that neatly connected Raspberry Pi 3B+ to the display. We tested it with a Raspberry Pi 4 using a Mini-HDMI to HDMI cable and it worked just fine.

Setting up

Our test unit came with a Raspberry Pi 3B+ and microSD card, along with some printed instructions. It was easy to disassemble and, consequently, reassemble. Just attach it to the GPIO pins, insert the HDMI and USB cables, and away you go.

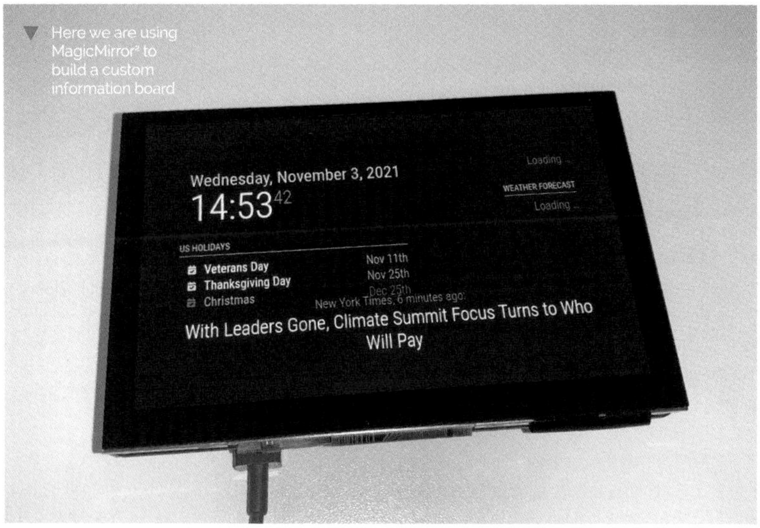

▼ Here we are using MagicMirror^2 to build a custom information board

config.txt

> Language: **Bash**

```bash
### Select Custom Timings
hdmi_group=2
hdmi_mode=87

## Disable audio over HDMI
hdmi_drive=1

# HDMI Timings for MCT050HDMI-A series
(800x480)
hdmi_cvt=800 480 60 6 0 0 0
```

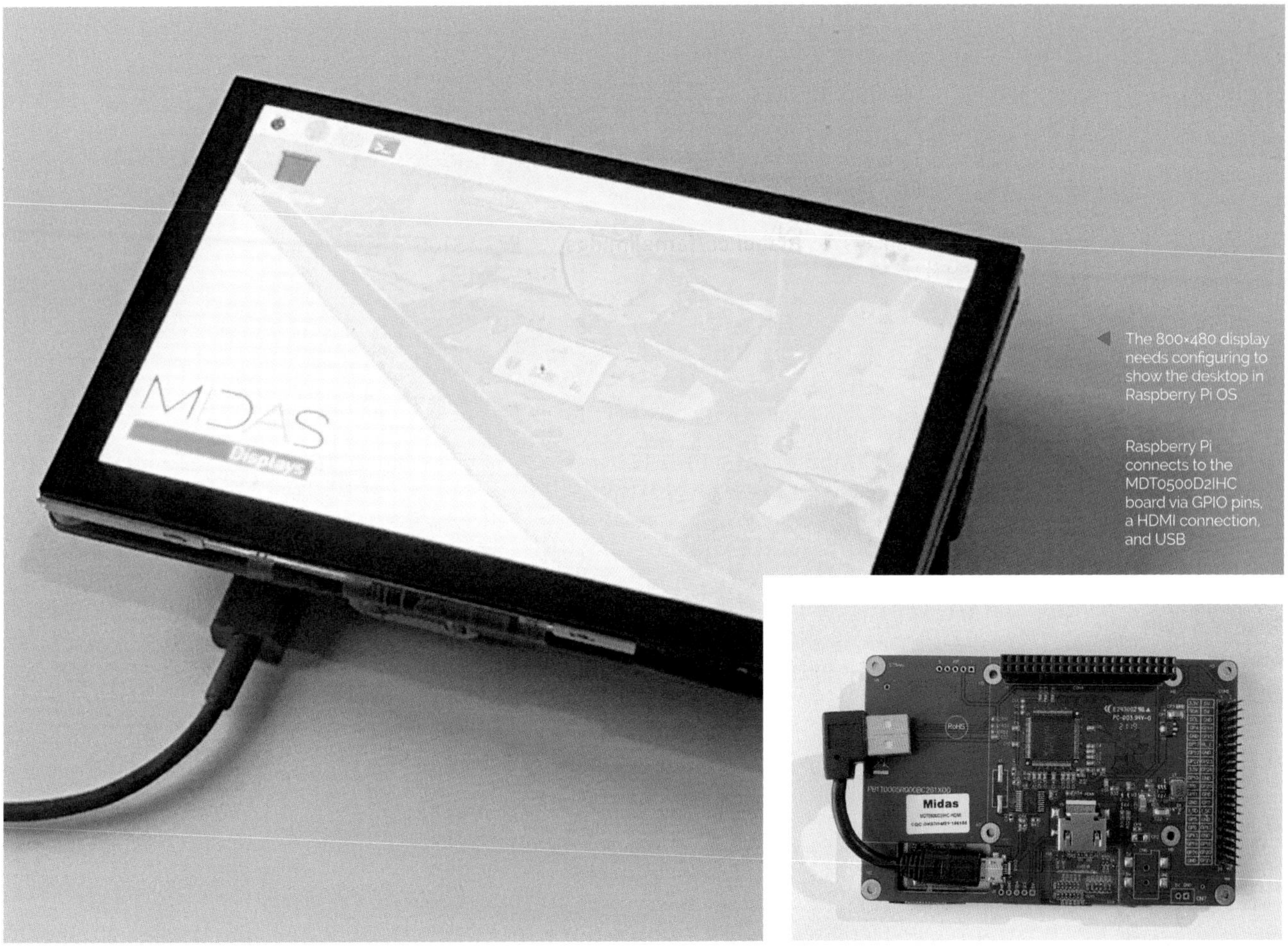

The 800×480 display needs configuring to show the desktop in Raspberry Pi OS

Raspberry Pi connects to the MDT0500D2IHC board via GPIO pins, a HDMI connection, and USB

The 800×480 resolution is non-standard, so getting the screen to work requires editing the **config.txt** file. It's nothing particularly hairy, but you will need to set custom timings. You can find the info you need on Raspberry Pi's Documentation site (**magpi.cc/custommode**). The key part is the timings, which you'll need to set at: 'hdmi_cvt=800 480 60 6 0 0 0'. We've included the **config.txt** additions in this review for reference.

What you can do?

With everything working, we had a lot of fun turning our MDT0500D2IHC into a variety of different projects. We turned it into a magic mirror (**magpi.cc/magicmirror**) and created a weather information board; we then used Info-Beamer (**magpi.cc/infobeamer**) to turn it into a photo display board and play video clips. In addition, we added an on-screen keyboard to Raspberry Pi OS using both Florence and Matchbox (**magpi.cc/onscreenkeyboard**).

> "The visual fidelity is superb and the touchscreen response is immediate"

On the whole, this is a great display. The visual fidelity is superb and the touchscreen response is immediate. The physical setup couldn't be any easier, and the configuration is easy once you have the correct timings info. While Midas provides a detailed datasheet (**magpi.cc/midas5indatasheet**), there is little tutorial support for beginners, presumably as a result of its industrial engineering background. But if you want a small screen to integrate into a build, or want to experiment with a small magic mirror or information screen; this is a great choice.

Verdict

A fantastic display with an excellent touchscreen and good Raspberry Pi integration. The configuration is easy enough once you have the correct timings. Midas could provide beginners with more documentation, though.

8/10

3.7" e-Paper **e-Ink Display for Raspberry Pi Pico**

▶ The Pi Hut ▶ **magpi.cc/epaperpico** ▶ From £24 / $28

Pico and e-ink should make the ideal low-power combination. By **Phil King**

SPECS

DISPLAY: 3.7-inch, 480×280 pixels, four-level greyscale

REFRESH: 3 seconds (full), 0.3 seconds (partial)

DIMENSIONS: 95.3×56.9 mm board, 47.32×81.12 mm display area

E-ink displays have long proven useful for portable Raspberry Pi applications with low power requirements where the display doesn't require rapid updates. Combined with the lower-power usage and tiny footprint of Raspberry Pi Pico, an e-ink display makes an ideal combination for projects like a name badge or weather display.

This 3.7-inch Waveshare e-ink display features two female headers on the rear so that you can simply plug in a Pico (equipped with soldered male headers). The graphic on the board makes it clear which way round to orientate Pico to avoid getting the wrong connections.

The board also features an eight-pin SPI interface, so can be wired up to a Raspberry Pi computer, Arduino, or another microcontroller.

Four shades of grey

While the 480×280 display is monochrome, it does offer four levels of greyscale for fairly accurate shading. One advantage of the greyscale display is that it takes less time to do a full refresh than on colour e-ink equivalents (which Waveshare also makes): just three seconds. Even better, it's possible to do a partial refresh of an area of the screen, for instance for updating a digital clock display, in a mere 0.3 seconds.

To make the e-ink display work, you'll need to visit the Waveshare wiki (**magpi.cc/epaperpicowiki**), and download a zip file containing C and Python example code and UF2 files to flash to Pico. The Python demo is limited and prints some text and a couple of rectangles in portrait mode. The C code – detailed in the wiki – uses landscape mode and is more extensive, printing text, drawing shapes, and showing a digital clock with seconds counting. There's no example of displaying a bitmap image, however, so you'll need to perform a bit of internet trawling to work out how to do it.

▲ As with all e-ink displays, anything on screen will remain there, even with no power

> "It does offer four levels of greyscale for fairly accurate shading"

▼ The rear features female headers to insert a Pico, plus an SPI interface to connect other boards

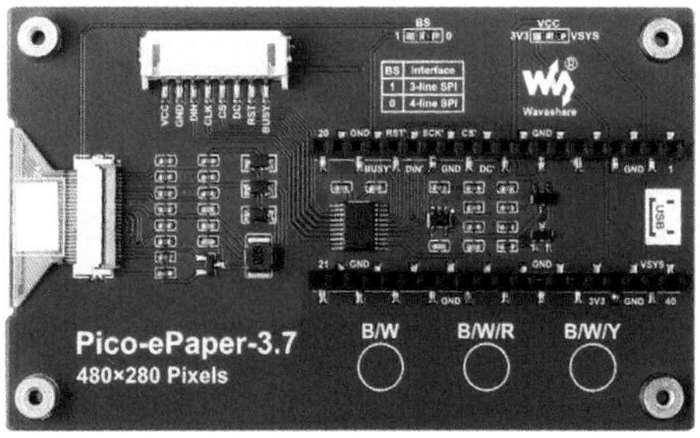

Verdict

The display itself works well, and is fast to refresh, but you'll need to do some work on the code to make the most of it.

7/10

Argon **IR Remote**

▶ Argon40 ▶ **magpi.cc/argonir** ▶ £9 / $10

Looking for a sleek minimalist remote for your Raspberry Pi media centre? **PJ Evans** sits back on his sofa and gets clicking

SPECS

DIMENSIONS: 150 mm × 35 mm × 10 mm

POWER: 2 × AAA batteries

COMMUNICATION: Infrared

HARDWARE COMPATIBILITY: Argon ONE & EON cases

SOFTWARE COMPATIBILITY: LibreELEC or Raspberry Pi OS

Using a Raspberry Pi 4 as a media centre makes for a brilliant way to access your favourite sounds and movies. Add Argon40's sleek ONE case and it looks as good, if not better, as any other set-top box. There's even mature software in the form of the LibreELEC operating system and Kodi media centre app. All that's missing is an easy way to control the action. The eagle-eyed may have noticed the ONE case's circuitry includes an IR sensor, and now Argon40 has made the companion Argon IR infrared remote control available.

Thankfully, Argon40 has been watching other tech design teams rather than those of most modern TVs. What you get for your very reasonable £9 is a minimalist, slim remote with just enough to make controlling Kodi a breeze. Basic directional control, selection, volume, and standard navigation work alongside a power button that works out-of-the box with any Raspberry Pi 4, regardless of operating system. It's light too, at just 65 g with two AAA batteries.

▲ The lack of extraneous buttons and a small form factor make the Argon IR comfortable to hold and use

More than media

If media centres are not your thing, or you have other epic plans for a remote control, Argon40 provides software that allows you to assign any key press to a button using the standard LIRC library with ease, making the addition of IR remote control to your project easier than ever before. In our tests, this worked flawlessly. However, we did have some issues configuring Kodi due to some missing steps in the supplied instructions. Once resolved, the remote performed admirably from across the room.

> A minimalist, slim remote with just enough to make controlling Kodi a breeze

▲ You'll need an Argon ONE case (or the forthcoming EON) to use the Argon IR

Verdict

Although the setup instructions could be improved, Argon40 has delivered a sleek, simple, and reasonably priced IR remote, both for Raspberry Pi media centres and homebrew projects.

9/10

10 Amazing: Displays

See what you're doing with these varied solutions. By **Rob Zwetsloot**

Some Raspberry Pi projects don't need a screen, and they're great. Sometimes, though, having something to see is a big part of a project. We don't just mean a standard monitor either: sometimes you need something a bit more specialist. Here are some great options to choose from.

▲ 1.3″ SPI Colour Round LCD Breakout

Your new watch

A tiny yet vibrant, circular colour display for your smallest Raspberry Pi projects, or at least the projects that need a 1.3-inch round screen.

£23 / $33 | **magpi.cc/roundlcd**

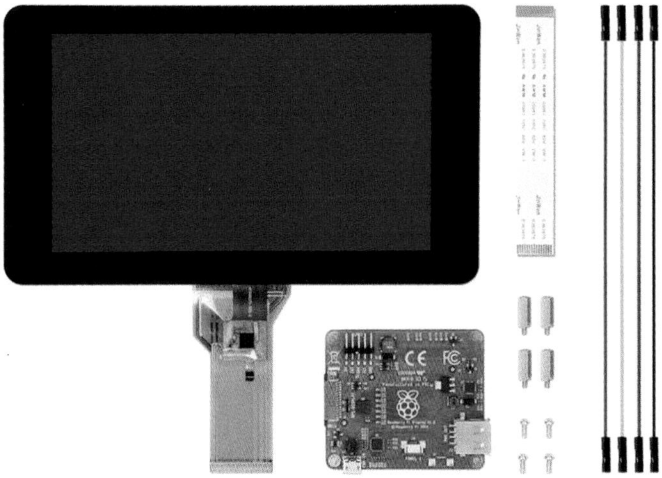

▲ Official Raspberry Pi 7″ Touchscreen Display

A great standard

The official display is a nice, fairly large screen that is used in many tablet projects, especially as it's touchscreen as well. It's 800×480 pixels and connects to the DSI Display port on Raspberry Pi.

£60 / $85 | **magpi.cc/officialdisplay**

▸ Inky Impression

Living paper

This 5.7-inch, seven-colour e-paper display is gorgeous, and perfect for projects that don't need a high refresh rate or need to look a certain way.

£66 / $94
magpi.cc/inkyimpression

▲ HyperPixel 4.0

High-resolution display

This 800×480 display is super-reasonably priced, and also comes in a touch version. It's bright and vibrant and honestly one of the best smaller traditional screens for Raspberry Pi.

£39 / $55 | **magpi.cc/hyperpixel**

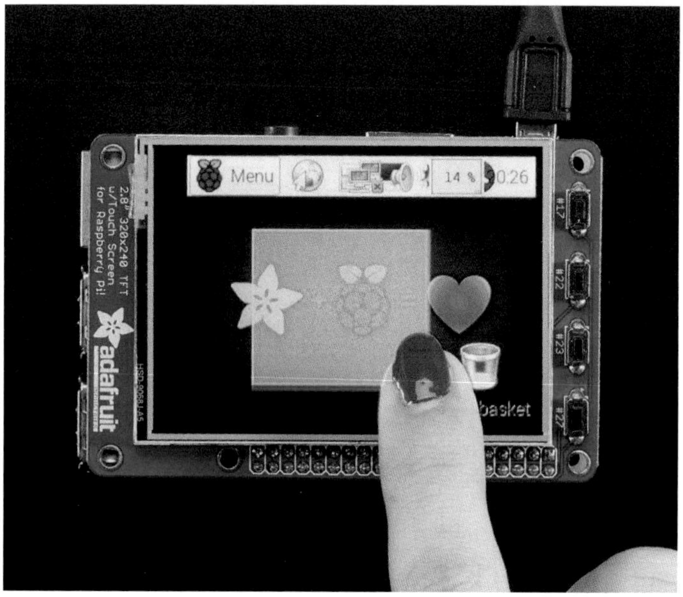

▲ PiTFT Plus

Classic display HAT

Probably the first big display add-on for Raspberry Pi, the PiTFT is still a great bit of kit. It has some physical buttons, along with touch sensitivity.

£41 / $35 | **magpi.cc/pitft**

▸ GFX HAT

Mechanical utility

A different style of display, the GFX HAT is a simpler screen that also includes nice capacitive touch buttons. Perfect for projects that need a readout and some interactivity, like a thermostat.

£23 / $32
magpi.cc/gfxhat

▲ A normal monitor

You probably have one

If you want to use Raspberry Pi as a desktop computer, there's no better way than a standard monitor. Just plug it in with an HDMI cable and off you go.

Prices vary

▲ 2.23-inch OLED display HAT

Minimalist and sharp

This small screen is OLED, making it very vivid, but also very simplistic, so it's it perfect for projects that need a basic and stylish read out.

£16 / $23 | **magpi.cc/oledhat**

▸ 1.54″ SPI Colour Square LCD Breakout

Four awesome sides

If a round display won't cut it, can we tempt you with its more angular sibling? It's slightly larger but still very small at 1.54 inches.

£18 / $26
magpi.cc/squarelcd

▲ 7″ HDMI LCD

Tablet monitor

An alternative to the official display, with higher resolution. It connects via HDMI, so it's a bit of a more traditional monitor, although it does have its own case with a kick-stand.

£55 / $78 | **magpi.cc/hdmilcd**

10 Amazing: Wearable projects

Ways to wear your Raspberry Pi with pride

As Raspberry Pi is so small and draws so little power, it's a perfect device for powering more complex wearable projects. We've done cosplay stuff with it in the past, and here are some other incredible projects that let you wear your Raspberry Pi.

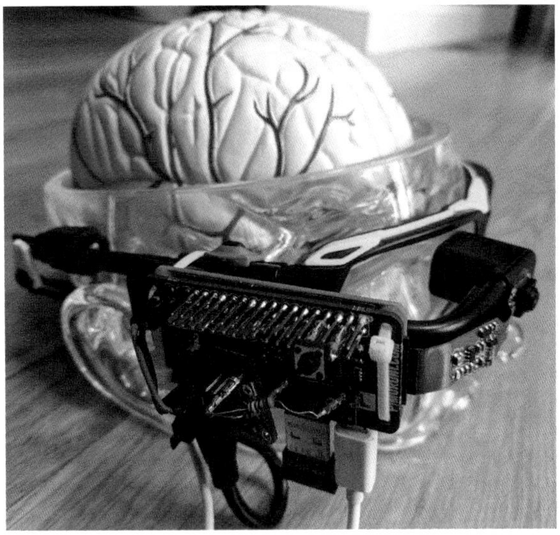

▲ PiGlass

Smart vision

We maintain that Google Glass was cool, so seeing excellent DIY versions using a Raspberry Pi is always a delight. This one packs in a lot of extra features as well.

magpi.cc/piglass

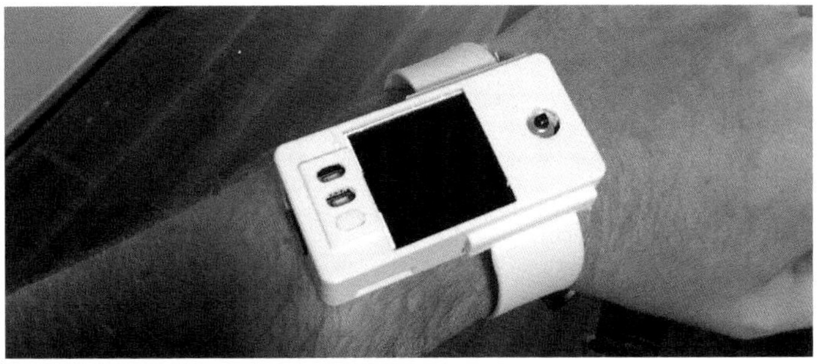

▲ Raspberry Pi Smart Watch

Wrist computer

With screens getting smaller and smaller, discreet and wrist-bound Raspberry Pi builds are easier than ever to do. This Smart Watch build is a very fun example.

magpi.cc/smartwatch

▸ Social media without the internet

Offline socialising

This art piece made full use of wearable tech by having interactive objects all over these coveralls. This allowed for social media-style interactions in real life.

magpi.cc/socialwear

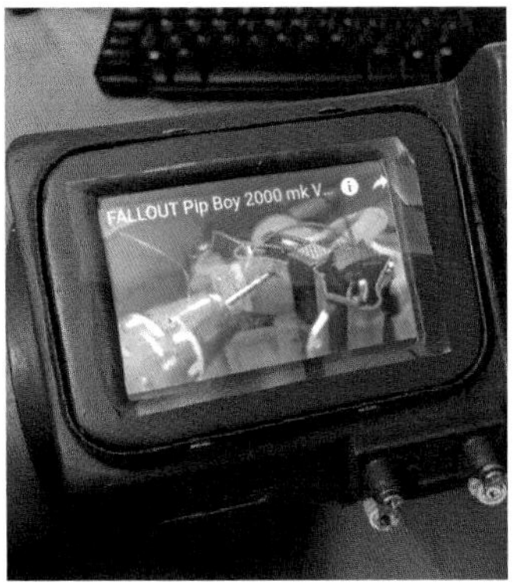

◂ Pip-Boy built from scrap

Apocalyptic wrist computer

You can buy official Pip-Boy cases, or 3D-print carefully crafted models. However, the recycling nature of this project makes it a firm favourite of ours.

magpi.cc/pipboyscrap

Wearable Tech Projects

Want to see more wearables, and maybe make some yourself? Check out *Wearable Tech Projects* by Sophy Wong and our sibling mag, HackSpace Magazine:

magpi.cc/wearableprojects

▲ PiE-Ink Name Badge

Who are you?

This is sure to impress folks when you're at an event (whenever they come back) or when you're the newbie in the (geeky) office.

magpi.cc/pieink

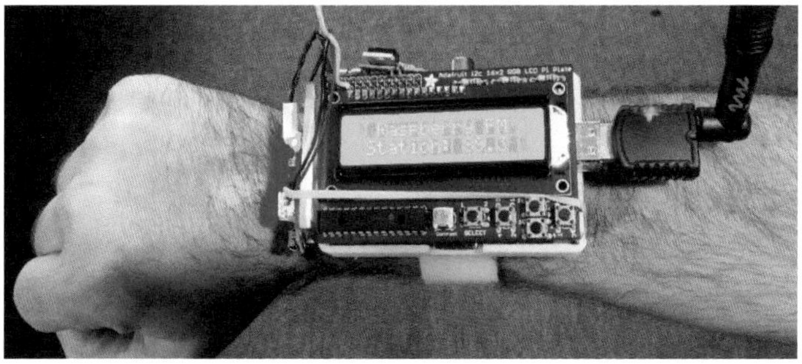

▲ RaspWristRadio

Portable FM

We like the very 1980s movie kid inventor look of this DIY radio that fits on your wrist. Don't be rude, though: bring headphones to listen!

magpi.cc/raspwristradio

▲ Wearable Cyberpunk Gesture Pad

Hacker wear

Need a wrist-mounted touchpad with multi-touch gestures? Then look no further than Zack's cyberpunk-inspired gesture pad.

magpi.cc/gesturepad

▲ Wearable Time-Lapse Camera

Taking a walk

This Raspberry Pi Zero project is simple yet cool, and we think it would make for some great music video footage.

magpi.cc/timelapsecam

▶ Smart Cap

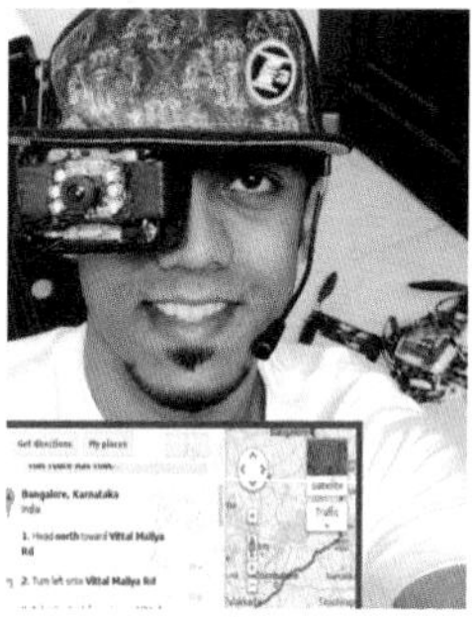

Cyber eye

This more conspicuous take on a pair of smart glasses can be used on multiple types of headwear. It's also completely open.

magpi.cc/smartcap

▲ Project Glass

Real-life subtitles

Real-time translation, while not 100% accurate, is quite impressive. Using this AR-like Raspberry Pi glasses system, you can use it to try to talk to real people.

magpi.cc/projectglass

10 Amazing: Raspberry Pi instruments

Get musical with these Raspberry Pi-powered musical creations

Raspberry Pi is an incredible piece of hardware for audio, and we know many people that use it for amazing sound systems. Did you know you can also create music on it? Not just with code either, as these projects show!

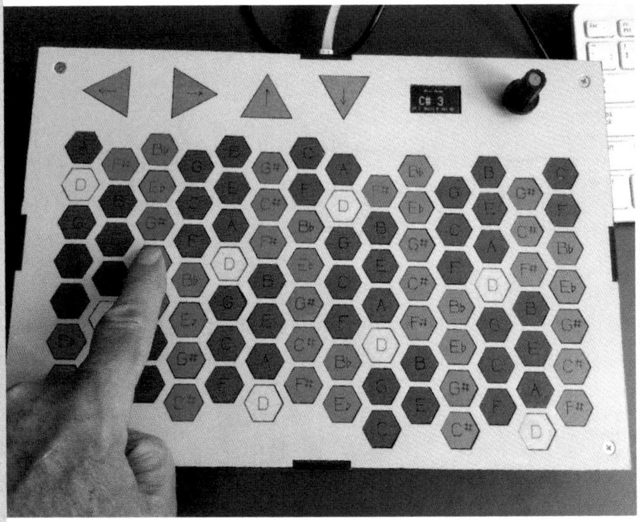

▲ Hex Keys

Isomorphic keyboard

One of the great instruments designed by Mike Cook, this laser-cut keyboard is quite different to a traditional piano keyboard, but can still make amazing tunes.

magpi.cc/pibakery

▲ Soundfighter

A new challenger

What do you do once you've mapped piano keys to a PS2 controller? You build a second, identical piano and duke it out in Street Fighter. Go home and be a piano man.

magpi.cc/soundfighter

▸ Piano stairs

Up the scale

You don't just have to make music with your hands – with a little inspiration from *Big*, you too can turn a little fun stair workout into a musical experience.

magpi.cc/pianostairs

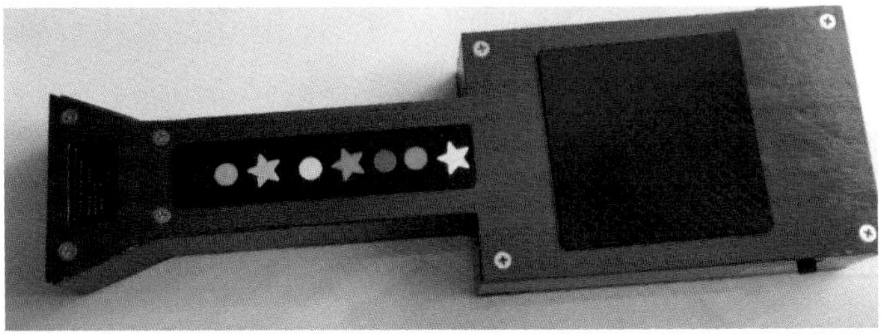

▲ Trill MIDI Guitar

New electric guitar

Another invention from Mike Cook, this excellent MIDI guitar uses special Trill touch sensors and a little display to create a cool, customisable guitar.

magpi.cc/trillguitar

▲ LED piano visualiser

Sound to light

These kind of visualisers are really popular on YouTube for pianos, giving them a rhythm game vibe. Create one for your own piano.

magpi.cc/pianovisualiser

▸ Pedal Pi

Effects via Raspberry Pi

This amazing project isn't an instrument per se, but it is part of one – a guitar effects pedal with a huge amount of customisation.

magpi.cc/pedalpi

▲ Self-playing xylophone

Robotic glockenspiel

With as few wires as possible, Stéphane was able to create this automated xylophone that uses many solenoids to play out a MIDI file on the classic instrument.

magpi.cc/xylophone

▲ MIDI Trumpet Robot

Automated jazz

This trumpet is played by robotic fingers depressing the valves, with MIDI sound actually playing from a speaker at the mouthpiece. It's controlled over the web.

magpi.cc/miditrumpet

▸ Ultrasonic theremin

Spooky tunes

With an ultrasonic distance sensor, you can control notes on Sonic Pi to create eerie tunes, perfect for a 1950s alien invasion B-movie vibe.

magpi.cc/theremin

▸ Raspberry Pi Guitar Rig

One-man band

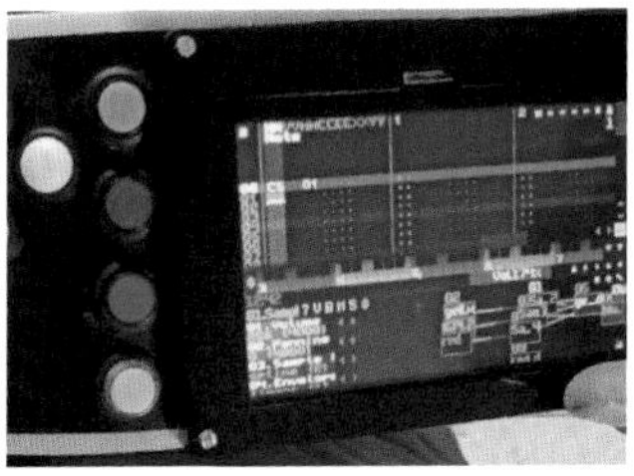

Building an amp into an acoustic guitar was only the first step. Adding a Raspberry Pi and touchscreen to control it on the fly to change synthesizer settings makes this extra-special.

magpi.cc/guitarrig

sinclair

the

COMPUTERS

THAT MADE

BRITAIN

"The Computers That Made Britain is one of the best things I've read this year. It's an incredible story of eccentrics and oddballs, geniuses and madmen, and one that will have you pining for a future that could have been. It's utterly astonishing!"

- **Stuart Turton**, bestselling author and journalist

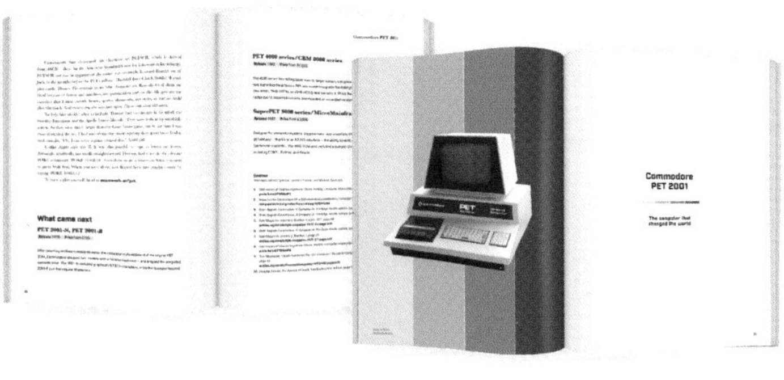

OUT NOW

Buy online: **wfmag.cc/ctmb**

Available on amazon